A Naturalist's Guide to Forest Plants

By the same author

A Naturalist's Guide to Field Plants: An Ecology for Eastern North America

A Naturalist's Guide to Seashore Plants: An Ecology for Eastern North America

A Naturalist's Guide to Wetland Plants: An Ecology for Eastern North America

A Naturalist's Guide to
Forest Plants

An Ecology for Eastern North America

Donald D. Cox
Illustrations by Shirley A. Peron

Syracuse University Press

Copyright © 2003 by Syracuse University Press
Syracuse, New York 13244–5160

All Rights Reserved

First Edition 2003
03 04 05 06 07 08   6 5 4 3 2 1

The paper used in this publication meets the minimum requirements
of American National Standard for Information Sciences—Permanence
of Paper for Printed Library Materials, ANSI Z39.48–1984.∞™

Library of Congress Cataloging-in-Publication Data

Cox, Donald D.
A naturalist's guide to forest plants : an ecology for eastern North
America / Donald D. Cox ; illustrations by Shirley A. Peron.— 1st ed.
p. cm.
Includes bibliographical references (p.).
ISBN 0–8156–0779–2 (pbk. : alk. paper)
1. Forest plants—Ecology—East (U.S.) 2. Forest
plants—Ecology—North America. 3. Forest ecology—East (U.S.) 4.
Forest ecology—North America. I. Title.
QK115.C7198 2003
581.7'3'0974—dc21
2003006884

Manufactured in the United States of America

To Barbara,

my wife and loving companion of forty years

Donald D. Cox studied at Marshall University, received a Ph.D. from Syracuse University, and was a professor of biology for forty-one years. His publications include *Some Postglacial Forests in Central and Western New York State; The Context of Biological Education: The Case for Change; Common Flowering Plants of the Northeast; Seaway Trail Wildguide to Natural History;* and from Syracuse University Press, *A Naturalist's Guide to Wetland Plants: An Ecology for Eastern North America* and *A Naturalist's Guide to Seashore Plants: An Ecology for Eastern North America.*

Contents

Illustrations

Table

Acknowledgments

I am indebted to Dr. Kenneth Heilman for reading the section on poisonous plants. I wish to thank Barbara Cox, Shirley Peron, and Robert Salisbury for providing many helpful suggestions. I especially wish to thank Sharon Doerr for carefully reading the manuscript and offering suggestions for improvement. I am grateful to Mark Lattime for graphic assistance and to Cindi and Kurt Mangione for technical assistance. I am grateful for the courtesy and cooperation from the staff of Penfield Library at SUNY-Oswego. My thanks to the staff at Syracuse University Press for creative suggestions and high professional standards. Finally, I want to thank Lisa, Allison, and David for support and inspiration during the preparation of the manuscript.

For the botanical and common names of plants, I have, when applicable, used those presented in the second edition (1991) of the *Vascular Plants of Northeastern United States and Adjacent Canada* by Henry A. Gleason and Arthur Cronquist. For plants growing outside the range of the above work, I have referred to the *Manual of Southeastern Flora* by John Kunkle Small.

Introduction

The importance of plants to all life on earth cannot be overemphasized. Without them all animal life would soon disappear. Plants are at the base of all food chains, and they also produce the oxygen essential for most living things. In addition to supplying food and oxygen, plants are an indispensable element of many other aspects of human civilization. Their presence ranges from the plastic components of computers to book pages like the one on which these words are printed. For the professional botanist, the farmer, and those who sell plants or plant parts, they are a source of income. For naturalists and others who love nature, they are a source of pleasure. For all of humanity they are a necessity.

Plants are all around us. We see ornamental plants in waiting rooms, offices, shopping malls, and many homes. City planners plant trees along their streets. In spite of this familiarity, or perhaps as a result of it, plants are frequently taken for granted. They are sometimes perceived as being less alive because they do not move about like animals. Yet each plant species has a unique set of features that enable it to survive in its environment. It is the aim of this book to raise the awareness of readers so they can appreciate forest plant species as successful living entities and for their importance to humans and the ecology of the earth.

In the following chapters, for ease of reading, technical terminology has been kept to a minimum. Some terms used to describe plants, although not highly technical, may have special meanings that are unfamiliar to the reader. For these, a glossary has been included. As one's interest in and knowledge of plants grow, invariably a point is reached where using common names to describe species is no longer satisfactory. Relying on common names can be confusing because every region may have a different name for the same plant. For this reason, the first time a common name is used in each chapter, the botanical name is given in parentheses. With a little practice these names will become as easy to use as common names, and they are much more reliable. The botanical name for a species is the same all over the world.

The way forests function as ecosystems and their role in the ecology of the earth are discussed in chapter 1. In chapter 2, brief descriptions are given for the major types of plant groups that grow in forests. The remainder of the book deals mainly with one group, the flowering plants or angiosperms. Emphasis has been placed on field observations, the intention being to provide descriptions and drawings that will aid in identification. An observer can learn to recognize the specific characteristics of many species with a little effort. In a given region of North America, there may be several thousand species of flowering plants. Trying to learn all of these is a daunting challenge, and the beginner may wish to begin by learning the plants in a limited group. For example, he or she could start by learning the trees, shrubs, toxic, medicinal, or edible plants in a woodland. All of these groups and others are described in the following chapters.

Forest plants have special characteristics that enable them to survive the woodland habitat. Those are described in chapter 3. In chapters 4, 5, and 6, soils, tree growth habits, and seasonal changes in forests are described. Chapter 7 presents toxic, medicinal, and edible plants of woodlands. Chapter 8 details inexpensive methods for those interested in collecting and preserving plants. Chapter 9 is offered for those who wish to go a step beyond identifying or collecting plants. It includes activities, projects, investigations, and thought stimulators.

Observing and learning about forests can be entertaining and educational. Since plants cannot run away, they can be observed or studied in detail in their natural habitats. While this is convenient for all who study or observe them, it makes plants vulnerable to all sorts of destructive forces. Their greatest threat is disruption of habitats resulting from human activities. The destruction of natural areas is increasing as the human population grows, making habitat conservation an urgent priority. These and many other topics are discussed in the chapters ahead.

The ability to identify plants has its own reward in personal satisfaction. Recognizing some of the local plants species when visiting a new woodland is like seeing old friends. It is comforting to know that even in a strange environment there are familiar "faces." In addition, as one travels through the countryside, being aware of forest communities that make up the landscape gives it a richer meaning and enhances the enjoyment of viewing it. It is the aim of this book to give the reader a broader understanding and a greater appreciation for woodlands and forest plants.

A Naturalist's Guide to Forest Plants

1

The Forest as an Ecosystem

Forest Origins

The origin of forests begins with the origin of land plants. The most radical change that ever happened to plant life on earth was the transition from water to a land environment. The first land plants evolved from green algae living on the ocean shore in the area between low and high tide, the intertidal zone, where they had daily periods of exposure to air. This was a time when geological forces within the earth were causing the intertidal zone to be pushed upward. As it was elevated, tidal pools formed; with further lifting, even these disappeared. During the millions of years this process took, those species who could adapt to the increasingly dry environment in which they found themselves became land plants. Many species whose rate of evolution could not keep pace with the rate of shoreline elevation became extinct.

The fossil record indicates that the first land plants may have resembled low-growing, branched, green sticks. Later the ends of the branches flattened and fused to form leaves. The plants that survived the rigors of natural selection had several structures that are necessary for survival on land. One was a root system that anchored the plant to the soil and absorbed water and mineral nutrients. Another was a waterproof, waxy covering, the cuticle, on aboveground parts, which prevented death from excessive water loss. A third feature was a series of tiny openings in the cuticle, the stomata, that allowed an exchange of carbon dioxide and oxygen during photosynthesis (fig. 1.1). Finally, survival on land required a conductive system to

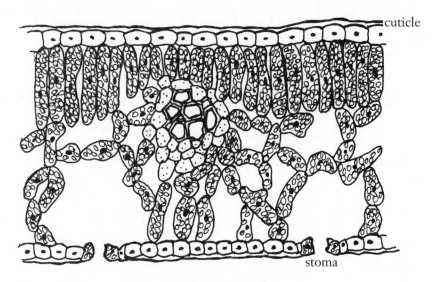

cuticle

stoma

1.1. Cross section of leaf showing stomata and cuticle

transport water and mineral nutrients from the roots upward and food material from photosynthetic tissue downward to the roots.

Mesophytes

Plants made the transition to land in the moist tropics. When compared on an environmental gradient ranging from all water to no water, this environment was in between. It was the type of habitat that supported mesophytic vegetation. Thus the first land plants were mesophytes, and most of the plants on earth today remain mesophytes. Typical mesophytic plants are found in areas where there is enough precipitation to support forest vegetation.

The greatest water loss by mesophytes occurs by evaporation from leaves, or transpiration. More than 90 percent of this loss is by evaporation through the stomata, and the rest is lost through the cuticle. Normally the stomata are open during daylight hours, allowing carbon dioxide to diffuse into the leaf. Carbon dioxide is one of the essential raw materials for photosynthesis. However, when the stomata are open, water molecules can diffuse out of the leaf. This loss of water is an unavoidable side effect since photosynthesis is essential for survival. In an environment where water is plentiful, as in mesophytic habitats, it does not pose a serious threat to the plant.

Most of the water loss of mesophytes takes place during the day. The cuticle is not completely waterproof, so some water continues to be lost during the day and night. Probably less water is lost at night because the temperature is usually lower. The greatest avenue of water loss, the stomata, are closed at night.

Xerophytes

Plants that are able to survive in environments where there is very little water are called xerophytes (pronounced zerofites). They have specialized structures or growth habits that enable them to tolerate or avoid long periods of drought. Their water conservation measures are more efficient than those of mesophytes. For example, a greatly thickened cuticle reduces water loss by evaporation. In some desert plants, the stomata are closed during the day, when water loss would be greatest, and open at night. Special adaptations allow these plants to store carbon dioxide in chemical compounds at night when it is absorbed, then release it for photosynthesis during the day. Xerophytes evolved from mesophytes in response to increasingly dry climatic conditions.

Even in the eastern deciduous forest, where precipitation is 30 inches (75 cm) or more per year, there are locations that are periodically deprived of water. For example, rock surfaces, tree trunks, fences, and so on are wet only during rains. Lichens and some mosses grow in these places, and they are efficient xerophytes. During extended periods when there is no rainfall, these plants dry out and give the appearance of being dead. When moisture is again available, they rehydrate and continue to photosynthesize.

Hydrophytes

Hydrophytes are plants that grow completely submerged, partially submerged, or with their roots in soil that is saturated with water for a portion of each year. These are discussed further in *A Naturalist's Guide to Wetland Plants* published by Syracuse University Press.

The Forest Ecosystem

Forests are essential members of complex relationships between living things and the earth called ecosystems. Forest plants are among the producers of the food that is the base for all food chains. Each species of plant performs a unique function in the environment. If a species becomes extinct, it leaves a hole in the ecosystem that cannot be filled by another. The smallest flowering plants, duckweeds, which may look like green pinheads floating on the water, are no less important than the tallest oak growing in the forest. They are each a part of a web of food and energy that affects many other species. The survival of both are necessary to maintain an undisturbed natural system. Without plants all animal life would soon disappear.

Forests are the most complex of all terrestrial ecosystems. They provide habitats for the greatest number of wildlife species in the natural world. They are the world's greatest reservoirs of diversity of wild living things. Forested areas are like great sponges that absorb rainfall and snow meltwater, then release it slowly to replenish streams and underground water sources or aquifers. In this way they regulate the flow of water so that it continues to be available during dry periods when the need is the greatest.

Woodlands furnish many products that are essential to modern living. Some of these will be discussed in later sections. Forests serve the public directly in a wide range of recreational activities such as hunting, fishing, hiking, camping, photography, and nature study. The U.S. Forest Service has discovered that in all but one of its nine administrative regions the recreational benefits of national forests are more valuable than timber, grazing, mining, or any other forest product. It may be that the most important value of all is the role of forests in maintaining the earth's ecosystem.

Forests and the Atmosphere

Oxygen

One environmental interaction shared by all humans and probably the one most often taken for granted is breathing. Yet if any one of life's essential requirements could be called the most important, it is the oxygen in the air.

Humans can survive for approximately a month without food and about a week without water but only a few minutes without oxygen. The early atmosphere of the earth contained almost no oxygen. Billions of years of photosynthesis has resulted in an atmospheric oxygen content today of about 21 percent. The concentration has probably been at this level throughout the evolution of mammals. It may be that a drop of only a few percent would threaten the survival of these organisms, including humans.

Organisms that require oxygen use it to convert the chemical energy of food into energy necessary to live, grow, and move. Although they do not move the way animals do, most forest plants grow throughout their lives and require energy to produce leaves, flowers, fruits, and seeds. The process of converting food into energy is called cellular respiration, and it takes place in all living things. Carbon dioxide is a by-product of cellular respiration. During daylight hours, plants use carbon dioxide in photosynthesis, generating more oxygen than they use in cellular respiration. The surplus diffuses into the atmosphere. In addition to that released in cellular respiration, they must absorb additional carbon dioxide from the atmosphere. In the dark, plants do not generate oxygen, and like most other living things they release carbon dioxide into their environment and absorb oxygen from it.

The only important source of oxygen on earth is from the process of photosynthesis. It is photosynthesis that keeps oxygen at its present level in the atmosphere. It has been estimated that up to 90 percent of the earth's oxygen comes from microscopic plants called plankton in the oceans of the world. The remaining 10 percent comes from terrestrial vegetation, mainly forests. Together these sources replenish one half of the oxygen in the air each year.

Carbon Dioxide

Carbon dioxide is removed from the atmosphere by photosynthesis and returned to it by cellular respiration and decay. These processes balance each other and have kept carbon dioxide at about the same concentration for thousands of years. It currently makes up about .03 percent of the atmosphere, but it has not always been at this level. During the time of the coal age, about 300 million years ago, there may have been several thousand times more carbon dioxide in the air than today. It was reduced to the present concentration when the coal age forests were converted into great deposits of coal, oil, and natural gas instead of decaying and releasing the carbon dioxide they had photosynthesized back into the atmosphere. These fossil fuels provide 90 percent of the energy used today by humans.

The atmospheric carbon dioxide/oxygen balance maintained by photosynthesis and cellular respiration/decay began to change with the invention of the steam engine and the initiation of the Industrial Revolution about 1800. Burning coal to fuel steam-operated machines began to add more carbon dioxide to the air than was being removed by photosynthesis. At first the accumulation was very slight, but it increased as the use of coal, oil, and natural gas escalated. The invention and use of the internal combustion engine resulted in a great boost in emissions. Since the early 1800s, the concentration of carbon dioxide in the air has increased by 25 percent. The greatest increase has been in the last fifty years. Since 1958 its concentration has increased by 10 percent. It is now at a level estimated to be the highest it has been in 130,000 years.

Carbon dioxide is added to the atmosphere by a number of modern activities. Electricity is the most common form of energy used in much of the world. In the United States and most of the rest of the world, 90 percent of the electricity is produced by burning fossil fuels. Automobiles add 300 to 400 pounds of carbon dioxide for each tank of gasoline used, and air travel adds millions of tons more each year. Deforestation is a world problem that increases the carbon dioxide in the air in two ways: (1) when trees are cut they are burned immediately or eventually they decay, and (2) the carbon dioxide they would have used in photosynthesis stays in the air. All of these activities are certain to increase as world population grows.

A change in world climate is a sure thing if the atmosphere continues to accumulate carbon dioxide. It is called a greenhouse gas because it absorbs heat escaping from the earth and radiates it back like the glass panes in a greenhouse. Consequently, the greater its concentration in the air, the warmer the world climate. The average global temperature today is 1° Fahrenheit (.55°C) warmer that it was one hundred years ago. The seven warmest years on record in more than one hundred years of record keeping have occurred since 1980.

Using the current rate of increase in atmospheric carbon dioxide, computer models predict the global temperature could increase by 4 to 9°

Fahrenheit (2–5°C) by the year 2050. This would have disastrous effects on the world. Some forested areas in North America would become grassland or desert. The grasslands of the United States and Canada would probably become too dry for farming. Eventual melting of the polar ice caps would flood much of Florida, New York City, Los Angeles, and other coastal cities of the world. In the geologic past, rapid climate changes have been frequently associated with mass extinction of plants and animals.

The solution to the problem is easy to perceive but may be impossible to achieve. The obvious solutions are to stop using so much fossil fuel and to plant more trees. Trying to persuade the world to reduce the use of fossil fuels is such a complicated problem, though, that there is currently no practical strategy. Trees are wonderful storehouses for carbon, but to balance carbon dioxide emissions for the next fifty years would require planting trees in an area larger than Australia. This is not likely to happen, but planting trees and reducing the rate of deforestation would solve other environmental problems as well as contributing to the reduction of carbon dioxide accumulation.

Water

Plants add water to the atmosphere by evaporation through tiny pores in their leaves (see fig. 1.1). This process, called transpiration, results in the loss of 98 percent of the water absorbed by plant roots. The amount of water lost in this manner is sometimes phenomenal. A beech tree 16 inches in diameter will lose 100 to 200 gallons of water on a typical summer day. An acre of beech-maple forest may transpire 500,000 gallons to the air in a single day.

Although the extent to which plants influence regional weather has been incompletely investigated, there is little doubt that forests influence rainfall. The rain forest of the Amazonian basin adds as much water to the atmosphere each year as is carried by the Amazon River. This rain forest is being rapidly destroyed to create farmland. If deforestation continues, it may be that rainfall could decrease to the point that it would cause problems not only for the remaining forest but for the intended cropland as well. Worldwide deforestation, in addition to other ill effects, may carry the double threat of increasing carbon dioxide emissions and decreasing regional precipitation.

Conservation

Habitat Destruction

Since colonial times, the population of North America has grown from thousands to several hundred million. During much of this time, the prevailing attitude of the immigrants to this continent toward forests and the environment in general has been governed by what ecologist Daniel Chiras has referred to as the frontier mentality. This attitude views natural resources as unlimited and existing solely for exploitation by humans. Humans are seen as masters of the natural world rather than as just another of its many species. Mastering the environment has become what amounts to an all-out assault with little concern for the consequences. There has been expanded deforestation and mining operations, unsustainable agricultural practices, and increased generation of electric power and manufacturing, all accompanied by extensive urbanization. These activities have altered the landscape, sometimes despoiling it, and have polluted the air and water. It is as if humans have declared war on the natural world, and even if that was not the intention, the effect is the same.

Habitat destruction is the main cause of extinction today. As a result of human activities, habitat destruction has been particularly intense in the past fifty years. Between 1850 and 1950, eighty-six species of plants are known to have become extinct. In comparison, in the twenty-one years between 1973 and 1994, one hundred species may have been lost to extinction. During the same time, about 750 became endangered, which means they are in immediate danger of extinction. Another 1,200 are threatened and likely to become endangered in the near future. This is not a problem restricted to the United States or North America. Of the more than 270,000 species of ferns, fern allies, and seed plants that are known in the world today, about 10 percent are estimated to be endangered, threatened, or rare.

Forest Wilderness Areas

There has been interest in maintaining wild areas in North America from as early as the mid-1800s. Naturalists such as John Muir, Aldo Leopold, and Robert Marshall have argued for the establishment of wilderness areas that would be protected from the influence of civilization. Such areas are more important today than ever because in competition with humans for space, wild plants and animals are losing on all fronts. Each species that is lost to extinction represents the loss of a potential life-saving drug, a source of food, or the loss of genetic material that could improve a crop plant. We still do not understand completely how a forest ecosystem works. It may be that wilderness areas will provide the natural laboratories that will yield information relevant to human survival.

A wilderness area must be large enough to include a functioning segment of an ecosystem. For plants this means enough space to accommodate the life cycles of pollinators and for the operation of seed-dispersal mechanisms. The Wilderness Act of 1964 set 5,000 acres as the minimal size and required that the area show no signs of human activity. Areas preserved under this act are mainly in the western United States and Alaska. The Eastern Wilderness Act of 1974 allowed smaller tracts to be set aside and permitted some signs of early human endeavors. About half of the land preserved under this act is in the Florida Everglades National Park and the Boundary Waters Canoe Area in Minnesota.

In protected forest wilderness areas, humans are visitors who can enjoy the beauty and the solitude but cannot stay. Theoretically their impact should be no more than any other large animal in the preserve. In some regions, so many are finding wilderness areas attractive places to visit that their wildness is being threatened. They are being threatened also by air and water pollution. However, the greatest long-range threat to their survival may be global warming brought on by the greenhouse effect.

Uses of Forests

Ancient Forests

Forest historians have noted that as civilizations grow, forests always recede, and when societies fade, forests regenerate. Trees have provided fuel and building materials for every human culture that has arisen. Ancient humans would not have been able to move from their southern points of origin into colder climates without the heat from wood fires. The heat from wood fires made possible the smelting of ores and the transformation of clay into pottery. Tools made of wood enabled early humans to protect themselves and to cultivate food plants. Until fairly recent times, sailing ships made from wood were the chief methods of transporting goods over long distances. Before the use of electricity, mechanical power was provided by windmills and waterwheels, both constructed of wood.

Mesopotamia has been called the cradle of civilization. It was an ancient region located between the Tigris and Euphrates Rivers in what is known today as Iraq. More than five thousand years ago, a center of sophisticated culture developed in southern Mesopotamia called Sumer that relied heavily on cedar logs floated in rafts from northern mountains. As the trees were cut, the denuded soil eroded away, exposing sedimentary rocks that contained large quantities of salt. Over centuries of this practice, the streams filled with sediments that clogged irrigation ditches and, more importantly, raised the salt concentration in the soil of the food-producing fields. This resulted in a drop in food production that was a contributing factor in the fall of the Sumerian civilization. The excessive logging of the great forests that had contributed so much to the rise of this civilization brought on its destruction. It is hard to imagine the region as we know it today, barren and desertlike, instead covered with a great primeval forest that may have extended to the shores of the Mediterranean Sea.

This forest was the main source of energy and building materials for successive cultures for at least three thousand years. According to one forest historian (Perlin 1991), as early as 2650 B.C. the Phoenicians were supplying Egypt with cedar logs from this region that were almost 170 feet in length. Between 1000 B.C. and 900 B.C., during the reign of Solomon, an ancient King of Israel, thousands of Solomon's slaves, Israelites, and Phoenicians felled cedar trees for Solomon's building plans. This may have been one of the greatest clear-cutting projects in the history of forestry.

Phoenicia was a collection of city-states that occupied the region now known as Lebanon. The cedars of Lebanon *(Cedrus libani)* today have been reduced to a few hundred trees in an enclosure that tourists can visit. Lebanon is one of two nations in the world with trees on their flags (Canada is the other). The cedar on the flag of Lebanon is an icon of a heritage long lost to that country.

The Romans used phenomenal quantities of wood to maintain their society. During the first century A.D., public baths were very popular in Rome, and the city of probably a million people had dozens of large bathing establishments. A single modest bath required more than 114 tons of wood per year to heat it. In addition to home and public heating, Roman industries such as iron and copper smelting and glass and pottery making all required large amounts of wood. As the forests of the Roman peninsula were cleared, these industries moved into England and France where wood was still plentiful. Roman emperors financed the expansion and maintenance of the empire by mining silver ore in Spain and extracting the silver. Over a four-hundred-year period, the silver furnaces consumed more than seven thousand square miles (17,949 sq. km) of forest. Wood finally became so scarce that silver production dropped. This was one of the factors that contributed to the collapse of the Roman Empire.

By the time of the reign of King Henry VIII in England, the great forests of southern England, destroyed by the Roman iron smelting, had long since regenerated. But England had not kept pace with other European countries in maritime and industrial development, and the monarchy perceived this as a threat to national security. Iron smelting was initiated and the first products in 1543 were cannons. By the late 1540s, the iron smelting forges were using 117,000 cords of wood a year. A cord is a quantity of wood 4 feet wide, 4 feet high, and 8 feet long. For the next hundred years, smelting iron and copper, building ships, and making glass and charcoal as well as other traditional uses of wood resulted in the decimation of England's forests.

American Forests

The first use of American forests by England was in the late 1600s to supply wood to Barbados for construction and for use in the sugar industry. Barbados produced most of the sugar used in England, and one use of the wood was to make casks in which to ship the sugar. This drain on New England forests grew to 240,000 trees for the two-year period from 1771 to 1773. Also from the late 1600s, the Royal Navy considered American forests a valuable source of masts for ships. An act of Parliament forbade colonists from cutting white pine trees *(Pinus strobus)* larger than 24 inches in diameter 12 inches above the ground. These trees were named property of the crown and were reserved for use by the Royal Navy. This angered the colonists and was a contributing factor to the American war for independence.

When the first colonists arrived from Europe, most of eastern and western North America was covered by unbroken forests. The settlers came from farming cultures in a much-less-forested Europe. The trees were a hindrance to farming, so the forest was set upon with ax and saw to make fields for crops, meadows for grazing cattle, and lumber for building homes. As settlement expanded westward, more clearing was required for farming and more timber was needed for building. Some of the uses of wood were extravagant by today's standards. In the days before macadam and concrete road paving, some well-used roads, called plank roads, were covered by laying wood planks along the length of the road. The prevailing attitude was that the forests were unending and that they would last forever. Unfortunately, this attitude persisted, for many, into the twentieth century.

In the growth of America, the forests of the east were used first. The methods chosen to harvest trees often favored maximum profit at the expense of good logging practices. When logging was finished in an area, it was left with great quantities of brush, called slash, from trimming the logs. During dry periods, the slash served as kindling for forest fires. Many disastrous fires swept through northeastern forests from Maine and New Brunswick to Minnesota and Manitoba. These resulted in the deaths of thousands of people, wiped out towns and villages, and destroyed millions of acres of virgin forest. The fires were too numerous to name them all, but early ones were the Miramichi and Piscataquis fires of Maine and New Brunswick in 1825. These destroyed three million acres (1,209,677 hectares) of forest. A more recent fire burned Cloquet, Minnesota, to the ground in 1918, took 488 lives, and destroyed 255,000 acres (102,823 hectares) of virgin forest.

The loss from fires of human lives, property, and virgin forests is not the end of the story. The soil of areas that have been logged, especially those that have been burned over, is exposed to the forces of erosion. In the Great Lakes states that are drained by the Ohio and Mississippi Rivers, millions of tons of forest topsoil were carried into the Gulf of Mexico. Since it takes between two hundred and one thousand years to form one inch of topsoil, when the topsoil is gone, it is gone forever in terms of human life spans. Cut-over and burned-over forests provide few impediments to slow the runoff of rainwater, so even normal rainfall can cause flooding. Runoff from cut-over forests was a contributing factor to the great flood of 1937 in the Ohio and Mississippi river valleys. This flood took four hundred lives, left a million people homeless, and destroyed $5 million worth of property. When there is excessive precipitation, the floods are quicker and more severe in deforested areas than in those areas where the runoff comes from undisturbed forests.

The extensive old-growth or virgin forests that covered eastern and western North America when the colonists arrived are now gone. An old-growth forest is one that is hundreds or even thousands of years old and has never been cut over by humans. When the mature vegetation of an old-growth forest is removed by logging or burning, the process of ecological succession will eventually produce a secondary forest (see chapter 5). That is why forests are called renewable resources. Forests and forest products can last indefinitely as long as the amount of timber cut each year does not exceed the amount that is produced by normal growth. This amount is known as a sustained yield level of cutting. However, old-growth forests are considered nonrenewable for at least two reasons: (1) old-growth forests generally have a greater diversity of plant and animal species than do secondary forests, and (2) about seventy years after the initial cut, the trees are ready to harvest again. The forest is never given enough time to return to old-growth conditions. Thus when a virgin forest is cut down, it is lost forever.

Most of the forests in eastern North America have been cut over one or more times. Today, only a few isolated patches of old-growth trees can be found in eastern forests. One of these with fairly easy access by automobile is Hearts Content Sanctuary in the Allegheny National Forest near Warren, Pennsylvania. Twenty acres (8 hectares) have been preserved where one can observe the primeval white pine, hemlock, and hardwoods forest as seen by Native Americans and early settlers. This can be an unforgettable experience for those who cherish the forest ecosystem.

It has been estimated that in the lower forty-eight states, 95 to 97 percent of the original old-growth forests have been cut over. Most of the remaining virgin stands are in national forests of the Pacific Northwest. Timber removed from national forests can only be sold in the United States, but there is no such restriction on logs grown on privately owned forests. Lumber companies own great tracts of forest land in both eastern and western North America. Logs from these tracts are frequently sold in more-lucrative foreign markets. Then, to avoid a lumber shortage in the United States, there is continuing pressure on Congress to allow more cutting of old-growth timber in national forests. According to National Forest Service data, most of the national forests in Washington and Oregon have sold timber at a rate nearly 23 percent above the sustained yield level. If present cutting rates continue, according to one projection, the old-growth forests will be gone in fifteen to twenty years in the United States and in thirty to forty years in Canada. If future generations of North Americans are to have a timber resource, the practice of cutting more wood than the forest can produce must end.

Tropical Rain Forests

Tropical rain forests are broad-leaved evergreen woodlands with multiple-layered canopies that receive 100 or more inches of rainfall annually. The first forests to evolve on earth were tropical rain forests, and most of the plant and animal species on the planet still live there. Tropical forests are located roughly in the area between the Tropic of Capricorn to the south and the Tropic of Cancer to the north. These forests cover 6 to 7 percent of the total land surface, but they contain at least 50 percent and some estimates range upward to 90 percent of all species on earth. In addition to being the habitat of more plant and animal species than any other area on earth, tropical rain forests also provide many useful products for humans. One-half of the world's annual harvest of hardwoods is from tropical rain forests. Food products such as coffee, spices, nuts, chocolate, and tropical fruits and latex for the manufacture of rubber are tropical in origin. In addition, tropical rain forest plants are the source of life-saving drugs that are used to treat malaria, Hodgkin's disease, high blood pressure, heart disease, and other ailments. Of the three thousand plants identified by the National Cancer Institute as having compounds with potential for fighting cancer, three-fourths are rain forest plants.

The growth of the human population is threatening forests of the world as never before. This is especially true of tropical rain forests. In countries with rain forests within their borders, poverty has forced people to clear forest land to plant food crops and to cut trees for fuel wood. But most of the nutrients in rain forest ecosystems are in the vegetation, not the soil. When the forest is cleared, the soil is exhausted after a few crops and the farmer must move to another location, where the process is repeated. This is called slash-and-burn farming, and along with commercial logging it is rapidly destroying the world's greatest gene pool of biological diversity. Ecologist G. Tyler Miller Jr. has stated that we are losing 66,000 square miles (171,000 sq. km) of tropical forests each year. This is an area approximately the size of the state of Oklahoma. Even as this page is being read, a rain forest species may become extinct that carries a cure for cancer or some other deadly disease.

Tropical rain forests are located in three main geographic areas on the globe: Asia, Africa, and Latin America. Some tropical countries have already lost more than 95 percent of their forests. Estimates vary as to how long tropical forests can endure, but an optimistic one is that, at the 1998 rate of destruction, they will be gone in thirty years. Self-interest dictates that tropical forests be preserved for the material benefits they provide today and can provide for future generations. There is also a philosophical reason they deserve saving. The plants and animals of the tropical rain forest have been on earth longer and have as much right to be here as do humans. If one accepts this philosophy, it becomes our moral responsibility to protect other species from extinction.

The Firewood Crisis

More than half of all the timber cut in the world is used for cooking and heating. In less-developed countries, almost three-fourths of the inhabitants, who have no other source of energy, use wood to cook their meals. As early as 1985, nearly one-third of the earth's population was experiencing shortages of fuel wood. The United Nations Food and Agricultural Organization projected that by the year 2000 this number would approach one-half the earth's population. In areas of shortage, the countryside around villages is often stripped of everything that will burn. In southern Asia, Africa, and Latin America, forests are being destroyed, and bare soil has made erosion a national problem. Most of the countries where there are fuel wood shortages have poor forestry practices and tree-planting programs ten or twenty times smaller than necessary to meet future needs. The great numbers of people who are affected by this crisis, and the resulting damage to ecosystems, justifies urgent and immediate action by the United Nations.

Some Modern Uses of Forests

Most people know of the importance of plants for food but are unaware or take for granted the numerous other ways that plants enhance their lives. A few of the products that make forests indispensable in our society are wood for building and furniture, wood pulp for paper and plastic, and rubber for shoes and motor vehicle tires.

Wood

Lumber. The United States uses more wood per person than any other country in the world. Americans are surrounded by wood or wood products in all aspects of their lives. Most of the homes in North America are built on a framework of wood. Two important trees that produce lumber for these frames are Douglas fir *(Psedosuga menziesii)* and ponderosa pine *(Pinus ponderosa)*. Douglas fir is the most important timber tree in North America with ponderosa pine in second or third place. Both trees grow in the western forests that provide more than 50 percent of the timber used in the United States. The second largest source of lumber is the southeastern evergreen forest (see chapter 4).

Wood for many specialized uses is provided by eastern deciduous forest trees. Among these are beautiful furniture from walnut *(Juglans nigra)* and wild cherry *(Prunus serotina)*, baseball bats and tennis rackets from white ash *(Fraxinus americana)*, bowling pins and billiard cues from sugar maple *(Acer saccharum)*, barrel staves and hardwood floors from white oak *(Quercus alba)*, tool handles from hickory *(Carya spp.)*, and lumber for construction in water, such as boat piers, from bald cypress *(Taxodium distichum)*.

Throughout most of the twentieth century, the demand for wood in North America has been adequately supplied by native forests. As the population has grown, the demand for wood has increased accordingly. It has been predicted that at the current rate of use, the need will be 20 percent more than can be supplied by the year 2020.

Wood pulp. The cells that make up wood are relatively long and narrow and have walls composed mainly of cellulose. Pulp is made by treating wood chips with chemical and physical processes that reduce the chips to individual wood cells. Wood pulp is the raw material from which paper, synthetic fiber, and some plastics are made. North America leads the world in the production of pulp, most of which is used for making paper. The United States is the world's largest producer and user of paper. Enough paper is manufactured each year to provide each American with about seven hundred pounds. Compare this with India, where annual paper production is about five pounds per person. Soft wood such as that found in aspens, cottonwood, pines, hemlock, and Douglas fir are best for making paper. The southeastern pine forests are currently the main sources of pulpwood.

The huge quantity of paper used in the United States creates a great drain on forest resources and results in higher levels of air and water pollution. Increased use of recycled paper would improve both of these conditions. It has been estimated that if all the Sunday newspapers in the country were recycled, it would save about 500,000 trees each week. In recent years, between 25 percent and 30 percent of the waste paper in the United States has been recycled. Among the problems with recycling are that paper mills are not equipped with the machinery for removing inks and dyes from used paper, and it is less expensive to buy new pulp than to recycle. This will change as a growing population increases the demand for paper.

The first plastic was made by a Swiss scientist about the middle of the 1800s, but the modern age of plastics did not begin until after 1900 with the technology that produced rayon and cellophane. To make these substances, cellulose from wood pulp is dissolved and treated with other compounds. The resulting mixture is forced through tiny openings to form synthetic fibers that are woven into rayon fabric. If the mixture is pressed instead into thin sheets, it forms cellophane. Rayon was the first synthetic fiber, and, while it still has many uses, other synthetic fibers are available today. Among other products made from dissolved cellulose are photographic film, billiard balls, toys, and tool handles. Although millions of tons of wood pulp are used each year for the manufacture of dissolved cellulose plastics, most modern plastics are made from chemicals derived from coal, petroleum, and natural gas.

Latex

Latex is a milky fluid produced in the cells of certain plants. Among native plants, one source is the common milkweed (*Asclepias syriaca*). The latex of some tropical plants, when solidified, has elastic properties and will bounce if formed into a ball. The latex from other plants does not have this elasticity. Important commercial products of the these plants are rubber and chewing gum.

Rubber. The story of rubber is as full of adventure as that of the California gold rush. Early explorers took back to Europe stories of a strange substance used by tropical Native American cultures. In some of their games, they used balls that had a curious bouncing quality. When the substance was imported to Europe, it was looked upon as a curiosity for almost a hundred years. Joseph Priestly, the famous chemist who discovered oxygen, found that it could be used to rub out pencil marks and called it rubber. In 1823 a Scotsman named Charles Macintosh discovered that rubber dissolved in naphtha could be used to impregnate cloth that would then be water repellent. To this day in some parts of the world raincoats are still called macintoshes. In 1839 Charles Goodyear accidentally spilled a mixture of raw rubber and sulfur on a hot stove and discovered vulcanization. Unlike natural rubber, vulcanized rubber was not sticky in summer or brittle in winter.

The plant that supplies most of the world's latex for rubber is *Hevea brasiliensis*. It is a tree native to rain forests of Brazil. By the mid-1800s, a world demand for rubber stimulated the Brazilian government to pass laws against the exportation of rubber-tree seeds in an effort to keep its monopoly of this resource. Nevertheless, in the late 1800s seeds were smuggled out to the Kew Gardens in England where they were germinated. The seedlings were shipped to Ceylon (Sri Lanka) and became the nucleus for rubber plantations in Ceylon, Malaya, Java, and Sumatra.

During World War II, the Japanese occupied the islands in the southwest Pacific Ocean and stopped the flow of raw rubber to the rest of the world. In the mad scramble to find a substitute, chemists in the United States developed synthetic rubber. Today the amount of synthetic rubber produced approximately equals that of natural rubber. Our lives are surrounded by products made from rubber, many of which are made from a blend of natural and synthetic rubbers. At least two-thirds of the usage of rubber today is for motor vehicle tires. Modern radial tires require an increased proportion of natural rubber, and this has caused an elevated demand for raw rubber. Hoping to avoid a shortage in the future, the United States, Mexico, and Australia are exploring other sources of latex for rubber. A plant that shows promise is guayule *(Parthenium argentatum)*, a shrub that grows in the deserts of southwestern United States and northern Mexico.

Chicle. In the eyes of some, chewing gum is a colossal nuisance. Stepping on a wad of discarded gum on a warm summer day can indeed be a vexation. Nevertheless, gum chewing is a characteristically American pastime. The original chewing gum was made from chicle, a solidified form of the latex of the sapodilla tree *(Achras sapota)*. This tree grows in the rain forests of Central and South America, and its properties were known long before the first Europeans arrived in the Americas. Early explorers observed the Aztecs of the Yucatan peninsula chewing a substance that was later called chicle.

Chewing gum made from chicle first became available in the United States near the beginning of the twentieth century and rapidly grew in popularity. The demand soon outgrew the supply of latex from the sapodilla tree, and other sources had to be found. Today the chewing gum industry relies mainly on other gums or synthetics, but chicle is still an important base for some confections.

Human Evolution and Plants

Among living things, the genetic combinations that result in the greatest rate of survival occur more frequently in successive generations due to the pressures of natural selection. Humans evolved in close contact with the natural world around them, quite different from the conditions in which we live today. The results of natural selection were early human societies of hunter-gatherers: small bands made up of family groups who constantly wandered from place to place in search of new food sources. The genetic factors that distinguish humans from other animals have not changed in the intervening years, but the world of the hunter-gatherer has become one of steel, concrete, plastic, and machines. Although today some people have adjusted to this environment, many have not.

It is possible that humans are genetically wired to have a sense of well-being when in a natural setting away from the trappings of modern civilization. In *A Sand County Almanac*, Aldo Leopold stated, "There are some who can live without wild things and some who cannot." He identified himself as one who could not. He is not alone. American society is replete with evidence of the urge to return to nature. Millions of people each year visit national, state, county, and municipal parks. Camping is a favorite vacation activity for a large segment of the population. Public and privately owned campgrounds abound in every state in the nation. A recent guide listed at least 150,000 campgrounds in the United States. The most popular campgrounds are those that have lots of trees.

Other evidence in support of the concept that humans have a genetic need for nature is their extensive use of houseplants. This can be seen as an attempt to bring a bit of the natural world inside. Florists do a thriving business, and most homes in America have one or more carefully tended houseplants. Waiting rooms and many workplaces seem warmer and more friendly as a result of decorative plants. Even when, as in some instances, the plants are artificial, they still give a visual impression that is comforting. No shopping mall would be complete without a display of greenery often including medium-sized living trees growing on their concourses. Shade or ornamental trees are a normal landscaping component of lawns and parks. Whether or not they represent a genetic need, plants satisfy a very important aesthetic need in modern society.

2

Types of Plants

For many people, the word "plant" brings to mind a tree, a houseplant, a flower, or a weed. To be sure, these are all plants, but they are all of the same type. They all produce seeds. The seed plants are the ones we most often see because they are the largest and the most numerous plants on the earth. But there are other types of plants and it would be difficult to take a walk in the woods, or even in your backyard, without seeing some of them. This chapter explores the different forms of plant life that the naturalist is likely to observe in a forest.

Algae

The many different kinds of algae are classified by their color—blue-green, green, red, and brown—which indicates the presence of certain pigments and thus particular metabolic processes. They all have relatively simple structures, and none of them have roots, stems, leaves, or flowers. The algae have been on earth for more than three billion years, and those ancient algae were the ancestors of all modern plants. In woodlands the algae most likely to be seen are the green and the blue-green. The blue-green algae are less common than the green. They often appear as dark bluish-green, slimy patches on rocks or damp soil near water. These are among the oldest photosynthetic organisms on earth.

The blue-green algae are different from green algae in a number of ways. All algae have chlorophyll pigments, but the blue-green also contain a bluish-green pigment that gives them their characteristic color. They differ from green algae also in the arrangement of their genetic material. In green algae, as in most plants and animals, the genetic material or DNA in each cell is enclosed in a central structure called the nucleus. In blue-green algae, there are no nuclei: instead the strands of DNA are dispersed throughout the cells. In this regard, blue-green algae are more similar to bacteria than they are to green algae. Newer systems of classification reflect these similarities and place bacteria and blue-green algae in a separate group called the Kingdom Monera.

Green algae species are descendants of the blue-greens and are so called because the main pigments that give them their color are the green chlorophyll pigments. Green algae can be seen on tree trunks as green films on the bark. When seen under the microscope, this alga looks like clusters of tiny green marbles. The growth of green and blue-green algae in water reservoirs sometimes gives a bad odor and taste to drinking-water supplies.

Plantlike Organisms

Fungi

In older systems of classification, the fungi were included in the plant kingdom. This may have been because they lack animal characteristics more than that they possess plant features, but they have traits of both plants and animals. For example, they have rigid cell walls like plants, but like animals they do not possess chlorophyll or make their own food. However, the fungi are a very diverse group of organisms with characteristics that differ enough from both the plant and animal kingdoms to justify placing them in a separate category, the Kingdom Fungi. The ancestors of at least some of the fungi were probably green algae and, like the algae, they have been on earth three billion years or more.

The growth form of the fungi is basically filamentous, consisting of long, microscopic, threadlike strands. A large number of these strands, called the mycelium, are usually dispersed in the soil or in the dead body of a plant or animal. Most fungi are saprobes that obtain nourishment by secreting enzymes that digest organic material. At some point in the life cycle of many fungi, the strands of the mycelium grow together in a dense mass that appears above ground as a macroscopic fruiting body. The function of the fruiting body is to form microscopic reproductive cells called spores.

The spores are dispersed by wind, water, animals, and so on, and under adverse conditions they may remain viable for long periods of time. When they fall on a suitable medium, they germinate and grow into a new mycelium.

There are three major groups of fungi that the alert observer will be able to recognize in forests. These are sac fungi, club fungi, and slime molds.

Sac fungi (Ascomycetes). These fungi are very important to humans in several ways. On the dark side, they are the causative agents in such diseases as athlete's foot, ringworm, and ergot poisoning, sometimes referred to as St. Anthony's fire. They are also responsible for many serious diseases of crop plants. On the bright side, the yeasts used in baking and in brewing alcoholic beverages are sac fungi. Other species give the blue color to some cheeses and the distinctive flavor to Roquefort and Camembert, and sac fungi are the source of the antibiotic penicillin.

Some of these fungi have an aboveground fruiting body called an ascocarp. It is frequently shaped like a bowl or a concave disk. These range in size from less than $^1/_2$ inch to more than 5 inches (1–13 cm) in diameter. They can be observed throughout the summer in moist woods. The inside lining of the cup or disk may be a dull black or brown, or it may be brightly colored yellow, orange, red, or purple. This colored layer contains asci, or sacs, in which microscopic spores are produced that will grow into new underground mycelia.

Some sac fungi have ascocarps that are edible and delicious. Examples are truffles, which have never been successfully cultivated, and morels. Truffles are underground ascocarps that are famous in French cuisine. A common morel, morchella *(Morchella esculenta*, fig. 2.1), is 3 to 5 inches (7.5–12.5 cm) high and looks like a sponge on a stalk. They can usually be observed in woodlands for only about a month in the spring. They are sometimes mistakenly referred to as mushrooms, but true mushrooms are the fruiting bodies of club fungi.

Club fungi (Basidiomycetes). These are probably more familiar to the general population than any of the other groups of fungi. They are called club fungi because the microscopic structures that produce

2.1. *Morchella esculenta*

2.2. Gill Mushroom

spores, called basidia, are somewhat club-shaped. Two types of club fungi are commonly seen in the forest: gill fungi and pore fungi. Gill fungi are the common mushrooms of fields, lawns, and woods that grow in a great range of sizes and colors. A gill mushroom consists of a stalk and an umbrella-like cap (fig. 2.2). On the underside of the cap, thin sheet-like gills radiate out from the central stalk. The spore-bearing basidia are located on each side of the gills.

The color of the spores is an important feature in the identification of mushrooms. Spore color can be determined by making a spore print. The stalk is cut off near the cap, and the cap is placed, gill-side down, on white paper, and covered with a soup bowl or other convenient cover for a few hours. The spores will collect beneath the gills in lines the color of the spores. To proceed further with identification, refer to the references at the end of this book.

Pore fungi do not have gills. Instead the spore-bearing basidia are located in tiny tubes that open as pores on the underside of the fruiting body. These are best known as the bracket or shelf fungi that can be observed on any walk through the woods (fig. 2.3). Bracket fungi are important wood-rotting fungi that sometimes attack and kill living trees. Some of them cause dry rot and wood decay in houses and other wood structures. The fungal strands attack by digesting the cellulose in the wood. When the mycelium has become established in the wood, the shelflike fruiting body develops and airborne spores are dispersed. Some of the bracket fungi are thick and woody with white undersides that are sometimes used as surfaces for artwork.

Slime Molds (Myxomycetes). The slime molds are so different that they may not even be related to the other fungi. They are sometimes classified as be-

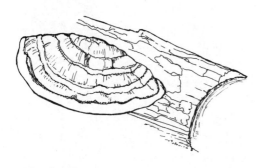

2.3. Bracket Fungus

longing to a separate kingdom, the Protoctista, which includes one-celled protozoans such as amoebas and paramecia. Instead of a filamentous structure, the growth form of slime molds is a macroscopic gelatinous mass of protoplasm called a plasmodium, with numerous embedded nuclei. Like a miniature version of a film monster, it flows very slowly along a surface such as a rotting log, engulfing tiny bits of organic matter that it digests as food. In one phase of its life cycle, it forms stalked, often brightly colored sporangia in which hard-walled spores develop (fig. 2.4). These are dispersed by wind and when one lands in a favorable environment it grows into a new plasmodium. Slime molds are most readily observed in moist places where decaying plant material is abundant.

2.4. Sporangium of Slime Mold

Lichens

Lichens are everywhere—on rocks, on the bark of trees, on fence posts, on the stones or bricks of buildings, and sometimes even on sidewalks. They are everywhere, that is, except in areas with heavy air pollution. In the centers of heavy industrialization where sulfur dioxide pollution is greatest, there are practically no lichens. This dead zone extends for a considerable distance, especially in the direction of prevailing winds. Lichens, then, or their absence, are indicators of air quality.

A lichen is actually not one organism, but two. It consists of an alga and a fungus living in very close association. This type of association in which there are benefits for both organisms is called mutualism. The alga may be a green or a blue-green, and the fungus is most often a sac fungus. The algal component manufactures food via photosynthesis while the fungus absorbs moisture and mineral nutrients. Most of the body of the lichen is made up of compactly interwoven fungal strands. The alga forms a very thin layer just below the surface and constitutes about 5 percent of the dry weight. This relationship of alga to fungus is a very complex one that required many millions of years to evolve. Fossil evidence indicates that lichens first appeared on earth in the Mesozoic era, which began about 225 million years ago.

Lichens are able to survive in very harsh environments. They grow on bare rocks where the temperature may reach 122°F (50°C) in summer and in the Antarctic where they survive and may even carry on photosynthesis at -65°F (-50°C). The surfaces on which they grow are usually nutrient poor, so their chief sources of mineral nutrients seem to be atmospheric dust and rainfall. Since nutrients are limited, the rates of growth are very slow, often less than one millimeter per year. Consequently, a large lichen is probably very old and some may be among the oldest living things on earth. A few Arctic lichens have been determined to be 4,500 years old.

Three major growth forms of lichens are easily observed in the forest: crustose, foliose, and fruticose. Crustose lichens are thin and very closely attached or embedded in the underlying surface (fig. 2.5). They can be seen on rocks but cannot be detected by touching. Some of the crustose forms are the most tolerant of air pollution. Because they may be present after other forms have disappeared from the area, they are indicator species of polluted air.

Foliose or leaflike lichens are thicker and usually have a central attachment to the substrate with unattached margins (fig. 2.5). During times of drought, the edges tend to roll up tightly and the lichen goes into a state of dormancy. With rainfall or an increase in humidity, it rehydrates and resumes photosynthesis. These and other forms of lichens are usually greenish-gray in color, but in alpine conditions and in the Arctic they may be bright yellow, orange, or red. Most foliose and fruticose lichens produce special reproductive structures called soredia that are dispersed by wind. Each soredium consists of a few fungal strands wrapped around one or more algal cells.

Fruticose lichens usually grow attached and perpendicular to the substrate, but some forms hang from tree limbs or other aerial perches. The fruticose forms are the most pollution sensitive of the lichens, and they are the first to disappear as air pollution increases. Familiar exam-

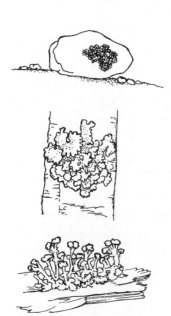

2.5. Crustose Lichen, Foliose Lichen, British Soldiers (Fruticose Lichen)

ples of these are reindeer lichen *(Cladonia rangiferina)*, British soldiers *(C. cristatella*, fig. 2.5), and old man's beard *(Usnea spp.)*. Reindeer lichens are common in the temperate zone on sterile acid soil of woody bog margins, but they are also an important food source for Arctic animals such as caribou, musk ox, and reindeer. British soldiers are normally about one inch high with bright red tops. Old man's beard is typically found in the southeastern United States in areas where the average relative humidity does not usually drop below 47 percent.

Mosses and Liverworts

Mosses and liverworts are collectively known as bryophytes. They are all small green plants whose ancestors were green algae. Mosses and liverworts are generally found in cool, moist habitats. They seem to be incompletely adapted to life on land because they lack many of the features that are necessary for survival out of water. For example, they do not have an efficient vascular system for absorbing and transporting water from the soil to all parts of the plant. For this reason, they are referred to as nonvascular plants. In complexity of the plant body and reproductive system, their evolutionary position is between algae and vascular plants such as ferns, gymnosperms, and flowering plants. However, bryophytes are not the ancestors of complex modern plants; they seem to be an independent line of evolution. The oldest known fossils of bryophytes are about 350 million years old, which is more recent than the oldest known fossils of vascular plants.

Bryophytes do not have true roots, stems, and leaves like the vascular plants. Instead of roots, they have threadlike structures called rhizoids that anchor them to the soil and function somewhat in the same way. Mosses and some liverworts appear to have stems and leaves, but they do not have the conductive tissue of vascular plants. Some of the liverworts are not differentiated into leaf and stemlike parts but consist individually of a flat ribbonlike structure called a thallus. It is usually fairly easy to distinguish between mosses and liverworts because mosses are erect and normally 1 to 3 inches (2.5–7.5 cm) high, while most liverworts grow flat against the ground. Individual bryophyte plants are small but typically grow in great masses on the forest floor. Plants of this group have a worldwide distribution but are more abundant in temperate and arctic regions than in the tropics.

The life cycle of bryophytes includes two genetically different phases.

The haploid phase has half the chromosome number, produces gametes, and is called the gametophyte generation. The diploid phase has the full chromosome complement, produces spores, and is called the sporophyte generation. Each of these phases is usually part of the plant body seen in the field.

The green leafy shoots of mosses are the gametophytes. These have sex organs at their tips that produce male and female gametes. When the sex organs mature during a rain or heavy dew, the motile sperm cell swims to and unites with an egg cell. The result of this union is a diploid cell, the zygote, which is the first cell of the sporophyte generation.

The diploid cell begins to divide immediately and forms a mass of cells embedded in the tissue of the gametophyte. These serve as the base for a long hairlike stalk that grows upward from the tip of the gametophyte. The stalk develops an enlarged capsule at its tip. The embedded tissue, the stalk, and the capsule make up the sporophyte.

In the field a moss plant is usually observed as a green leafy basal part, the gametophyte, with the green or brown stalk of the sporophyte growing from its tip (fig. 2.6). The capsule is usually bent downward and often has a pointed cap that is the remnant of the female reproductive organ. It is called the calyptra, and it is helpful in the identification of some mosses.

2.6. Moss Gametophyte and Sporophyte

Certain cells inside the capsule undergo a special kind of cell division called meiosis, or reduction division, in which the chromosome number is reduced by one-half. The haploid cells produced by this process are the spores that are the first cells of the gametophyte generation. The spores are dispersed by wind, and when one falls on moist soil, it germinates and grows into a green algal-like filament. Buds develop along the filament and each bud grows into a green moss gametophyte. Individual filaments are very difficult to see, but they may occur in networks made up of many strands. A diligent observer may be able to see them as green threadlike lines on moist soil near moss plants.

The mechanism for the dispersal of moss

spores may be seen with a simple hand lens. An intact moss plant complete with gametophyte and sporophyte can be collected and taken indoors. The capsule of the sporophyte has a cap that covers a ring of enfolded flaplike teeth. If the moss plant is placed under an incandescent lamp, the capsule will dry, the cap will pop off, and the teeth will flip outward expelling the spores. When the capsule is remoistened, the teeth will return to their enfolded position.

The life cycles of liverworts are very similar to those of the mosses except for the physical form of the sporophytes. In some of the thallose liverworts (those without leaflike and stemlike structures) such as marchantia (*Marchantia spp.*), the sporophytes are much smaller and hang downward from special upright branches that are reminiscent of miniature palm trees. In all instances, the sporophytes of the bryophytes are more or less parasitic on the gametophyte generation. Bryophytes are similar to vascular plants in exhibiting life cycles that alternate haploid and diploid generations. However, in all vascular plants the sporophyte is not parasitic on the gametophyte but is the most visible and dominant phase of the life cycle.

Ferns

The graceful beauty of ferns has always caught the fancy of humans, and almost everyone can recognize a fern. This group of plants is very diverse in size and growth form and many do not fit the mold of what is thought of as a typical fern. The water ferns are very small, but some of the tree ferns in the tropics may be more than 50 feet (15 m) high. Although they have a worldwide distribution in a wide variety of habitats from the equator to the arctic, 65 to 95 percent of all the fern species grow only in the moist tropics.

According to the fossil record, ferns have been on the earth for about 370 million years. Their ancestors were land plants that evolved from green algae, although all the in-between stages are not fully understood. The ferns are a large successful group of plants, but they have never been a major component of world vegetation. They grew abundantly on the floor of coal age forests as they do today in the redwood forests of California, the evergreen forests of the west, and the deciduous forests of the east. They have never been the most conspicuous or dominant plants, though, in any of these forests.

Ferns and other plants with specialized tubelike cells for conducting water are called vascular plants. Most of these, including the ferns, have well-developed roots, stems, and leaves. The ferns in the United States and Canada do not have erect aboveground stems but rather have underground ones called rhizomes. These often branch and radiate horizontally outward producing new clusters of leaves each year. The portion of the rhizome that grew the previous year usually dies, sometimes producing a circle of new fern leaf clusters. More commonly, individual clusters of leaves are observed in the field.

Fern leaves are called fronds. They consist of a stalk or stipe that is attached to the underground stem, with an expanded portion called the blade. In a few ferns the blade is undissected, forming a simple leaf. The walking fern (*Asplenium rhizophyllum*, fig. 3.1) and hart's tongue fern (*Phyllitis scolopendrium*) are species with simple leaves. In most ferns the blade is dissected one or more times, forming the lacy leaf that is typically associated with ferns.

The fern leaf grows in a unique manner. Embryonic leaves that develop on the rhizome are very tightly coiled. As they mature, they unroll from the base in a growth pattern known as circinate vernation. Because the young uncurling leaves look like the neck and head of a violin, they are called fiddleheads. The unrolling of fern fiddleheads is a welcome and attractive sign of spring. Fiddleheads of several species of ferns are collected in the spring and cooked as green vegetables. In some areas of Canada, fiddleheads of ostrich fern (*Matteuccia struthiopteris*) are harvested and frozen for commercial trade. These can be found in supermarkets in some American cities.

The life cycle of ferns includes a diploid sporophyte and a haploid gametophyte generation. Unlike the mosses and liverworts, the fern generations are completely independent of one another. The underground rhizome and the fronds make up the sporophyte generation. Species such as the cinnamon fern (*Osmunda cinnamomea*, fig. 2.7) and sensitive fern (*Onoclea sensibilis*, fig. 7.5) produce spores on separate stalks, but in most species, the spores develop on the undersides of the fronds.

The underside of a frond may appear to be covered with brown dots. These are fruit dots or sori, and each is a cluster of tiny, stalked, egg-shaped structures, the sporangia, that produce spores. Each sporangium contains special cells that undergo reduction division resulting in haploid spores with chromosome numbers half that of the sporophyte cells. Under drying conditions, the sporangium snaps open forcefully expelling the spores, which are scattered by air currents.

When a spore falls on moist, shady soil, it germinates and grows into a very small, green, flat, heart-shaped structure known as the pro-thallus. On the underside of the prothallus are rootlike outgrowths that attach it to the soil and organs for the production of male and fe-male sex cells. During a warm spring rain, a sperm cell swims to an egg cell, fertilizing it and form-ing a diploid cell that becomes the first cell of the sporophyte phase. This cell develops into a leaf that grows upward and a root that

2.7. Cinnamon Fern *(Osmunda cinnamomea)*

grows downward into the soil, and the new sporophyte is established. In ferns, the gametophyte and sporophyte generations are both green plants independent of one another.

In earlier times, before their life cycles were understood, ferns were considered to be mysterious plants. Observers assumed they reproduced by seeds, but no seeds could be found. The brown dots on the undersides of the leaves were observed, but no connection was made between these and fern reproduction, so the search continued for the illusive fern seeds. In those times, the unknown was often associated with magic. This may have given rise to the legend that anyone in possession of a fern seed was invisible.

Ferns are most often observed in woodlands, but they are frequently seen around the margins of wetlands. Most fern fronds die at the end of summer but some remain green all winter. They all produce new clusters of fronds each spring. Christmas fern *(Polystichum acrostichoides)* is a common attractive evergreen woodland fern. According to folklore, it is called Christmas fern because its leaf divisions resemble Christmas stockings hung from the fireplace mantel. Only one fern in the United States and Canada has the dubious distinction of being called a weed. The bracken fern *(Pteridium aquilinum*, fig. 7.6) was introduced from Asia and has become widespread in fields and open woods. It has a rapidly growing rhizome that sometimes becomes a nuisance in cultivated fields.

Fern Allies

Fern allies are the horsetails and clubmosses. They are called fern allies because they have life cycles similar to ferns. Like ferns, they are vascular plants whose ancestors evolved from the green algae, but they have actually been on the earth longer than ferns. The horsetails and clubmosses have independent sporophyte and gametophyte generations. The visible green plants in these groups are the sporophytes; the gametophytes are small or underground and are very rarely seen. Today the horsetails and clubmosses are small plants 4 inches (10 cm) to 3 feet (90 cm) in height, but their ancestors were the most important trees of the coal age forests. They attained heights of up to 60 feet (18 m) or more and grew in vast swampy forests more than 300 million years ago.

Horsetails

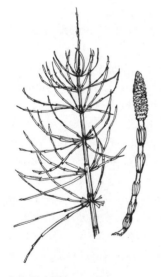

2.8. Field Horsetail
(Equisetum arvense)

The horsetails consist of a single genus, *Equisetum*, that is frequently found in moist or wet open habitats as well as in woodlands. In field horsetail *(Equisetum arvense)*, the sporophyte has two growth forms; a non-green spore-producing plant and a green photosynthetic one (fig. 2.8). The non-green cone-bearing branch arises from an underground rhizome in April. These have leafless, jointed stems, are whitish tan, and are 6 to 10 inches (15–25 cm) high. They shed their spores and wither in about two weeks. The green shoots appear a little later but do not reach their full development until late May or June.

The green phase of the field horsetail sporophyte consists of a central conspicuously jointed stem with branches in whorls. The jointed branches may have still smaller whorls of branchlets.

The whole plant presents a bushy appearance that reminded someone in the past of a horse's tail. The leaves are very tiny and nonfunctional; the green stem and branches carry on photosynthesis. This plant lives for one season and then dies back to the underground rhizome in autumn. The rhizome is perennial and puts up new spore-bearing and green shoots each spring.

Scouring rush *(Equisetum hyemale)* is another common species of horsetails. It has an unbranched, jointed, green stem topped by an egg-shaped cone (fig. 2.9). The stem is evergreen but new shoots appear each spring. The best development of cones is usually in June, but some searching in a colony of scouring rushes will turn up cones even in winter. Scouring rushes usually grow in very damp open woods or wet open spaces and are sometimes observed along railroad track clearings. The stems of both scouring rush and field horsetail contain a high amount of silica, the main ingredient in sand. In the past, they were used for scrubbing pots and pans, hence the name scouring rush.

2.9. Scouring Rush
(Equisetum hyemale)

Clubmosses

The clubmosses have been on the earth longer than either the horsetails or ferns. The name clubmoss is misleading because these plants are not related to mosses and resemble them only faintly. There are several genera, but the one most likely to be observed in the field is *Lycopodium*. Most lycopods have horizontal stems that grow over the surface or underground with erect branches that seldom exceed 8 to 10 inches (20–25 cm) in height. These stems may branch repeatedly, with older portions dying, giving rise to several independent clumps of plants. In this way large colonies can develop.

The lycopods are most often found in rich, moist woods. Shining clubmoss *(Lycopodium lucidulum)* has upright branches 4 to 6 inches (10–15 cm) high with densely crowded leaves (fig. 2.10). Spores are commonly formed from July to September in sporangia located in the axils of leaves near the stem tips. During this same period, small buds called gemmae develop at the stem tips. The buds may fall to the ground and grow into new plants.

2.10. Shining Clubmoss
(*Lycopodium lucidulum*)

They can be collected and cultivated in rich, moist forest soil.

In other clubmosses such as ground pine *(L. obscurum,* fig. 2.11) and ground cedar *(L. complanatum,* fig. 2.12), sporangia are clustered in cones at the tips of upright branches that usually shed spores in August and September. The spores are dispersed by air currents, and when they land on suitable soil they have a tendency to sift downward to a depth of 1 to 4 inches (2.5–10 cm). The spores germinate at this depth and the gametophytes develop underground. The gametophyte is non-green and obtains food through a mutualistic or parasitic relationship with a soil fungus. It varies in size and shape and may be up to an inch in diameter or length. Sex organs develop on the surface of the gametophyte, and in a film of water the sperm swims to the egg, fertilizing it and initiating a new sporophyte. The nature and location of sporangia, leaves, and horizontal stems are important features in the identification of the species of clubmosses.

Walking through a colony of clubmosses when the spores are mature

2.11. Ground Pine *(Lycopodium obscurum)*

2.12. Ground Cedar *(Lycopodium complanatum)*

will stir up clouds of yellow spores. They are produced in such great quantities that they are easy to collect. In colonial times, pioneer women used lycopodium spores as baby powder. Since they are water resistant, they have been used to coat pills to keep them from sticking together. Lycopodium spores are explosively flammable and have been used in fireworks and as flash powder for early photography. Clubmoss ancestors in the coal age also produced great quantities of spores. In fact, so many spores were released that in some regions, deposits of spores were converted to a special kind of coal called cannel coal. It is popular for fireplaces and campfires because it burns with a very bright flame and leaves only a small amount of ashes.

The clubmosses are evergreen. This is unfortunate for their survival chances because it has resulted in their being collected to make Christmas wreaths. In colonial times the practice may have been acceptable because the human population was concentrated along the eastern seaboard and numbered in the thousands. Today it is unacceptable because with a human population numbering hundreds of millions collection runs the risk of driving some species to extinction. The clubmosses are protected by law in several states and should be protected in all.

Gymnosperms (Conifers)

The gymnosperms are vascular plants that bear seeds in cones. The term gymnosperm means naked seed and refers to the fact that gymnosperm seeds are not enclosed in a fruit as are those of flowering plants. Gymnosperms have been on earth for about 325 million years. Their ancestors were vascular plants that evolved from green algae. They reached their peak of development during the time of the dinosaurs when they were the dominant vegetation on the earth. Since that time, they have declined as the flowering plants have become the dominant vegetation.

There are several groups of gymnosperms, but the two that are most likely to be observed in the United States and Canada are maidenhair trees *(Ginkgo biloba)* and conifers.

Maidenhair Tree

The maidenhair tree has leaves with the general shape of a frond of the maidenhair fern *(Adiantum pedatum)*. It is sometimes referred to as a living

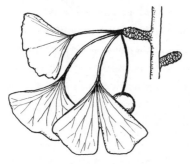

2.13. Maidenhair Tree *(Ginkgo biloba)*

fossil because it is the only surviving species of an ancient group of plants. The maidenhair tree is a native of China, but it probably does not grow wild anywhere in the world. It is a beautiful tree that is planted throughout the world for shade and as an ornamental.

The maidenhair tree has fan-shaped, yellowish-green leaves that turn a bright yellow in autumn (fig. 2.13). It has a remarkable resistance to smog and air pollution and has become popular for planting in cities. The dried, pressed leaves are sometimes used as bookmarks because they repel silverfish and other insects. There are male and female trees, but the male trees are favored for ornamental plantings. Male trees are preferred because the seeds of the female tree have a soft outer layer that decays and produces butyric acid, which has an offensive odor. Until recently it was impossible to distinguish between male and female trees until they reached seed-bearing age, but with modern technology the sex can be determined by identifying the X and Y chromosomes in very young seedlings.

Conifers

The most numerous and best-known gymnosperms are the conifers. These include such familiar trees as pine *(Pinus spp.,* fig. 2.14), hemlock *(Tsuga spp.,* fig. 6.11), fir *(Abies spp.),* douglas fir *(Pseudotsuga menziesii),* spruce *(Picea spp.,* fig. 2.15), and redwood *(Sequoia sempervirens).* Although most conifers, including all of the above, are evergreen, two genera in North America, larch *(Larix laricina)* and bald cypress *(Taxodium distichum),* shed their leaves in autumn. A few conifers such as some of the junipers *(Juniperus spp.)* and yew *(Taxus spp.)* are shrubby, but none are herbaceous. Included among the conifers are the world's tallest tree, a redwood 372 feet (112 m) high; a tree with the greatest circumference, a giant sequoia *(Sequoiadendron gigantea)* 84 feet (25 m) around the base; and the world's oldest living tree, bristlecone pine *(Pinus longaeva),* one of which has been determined to be 4,844 years old.

The leaves of the conifers are typically needle-shaped as in the pines, or they may appear to be overlapping scales as in arbor vitae *(Thuja occiden-*

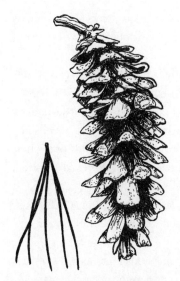

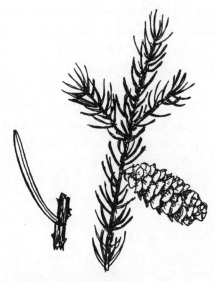

2.14. White Pine *(Pinus strobus)*

2.15. White Spruce *(Picea glauca)*

talis). In pines and larch most of the needles are attached in clusters on short spur branches. Pine needles are attached in clusters of two to five, while in larch the needles are usually shorter, softer, and attached in clusters of more than five. The needles of hemlock and fir are flat while those of spruce are four-angled and can be distinguished from other species by rolling them between the thumb and forefinger. Although a new crop of leaves is produced each spring in conifers, all of last year's leaves are not shed. Leaves remain on the tree for one to five years and are shed gradually throughout the year. The shape, type of attachment, length, and texture of the needles are all important features in identifying species of conifers.

In temperate North America, the conifers are very important economically as the main source of lumber for construction and forest products including pulp and paper. The forests of the Northwest supply the greatest quantity with douglas fir as the single most important tree in terms of the amount of lumber it provides. Although conifer forests have shrunk since the time of the dinosaurs, they still make up a major portion of the world's vegetation, covering vast areas of the earth's surface. For example, the southern coastal plain of the United States is characterized by a conifer forest of pines; the northwestern part of the United States is dominated by conifers; and the great coniferous boreal forest or taiga forms a band across Canada that extends through northern Europe and Siberia, making a ring around the top of the earth.

Like ferns and nonvascular plants, the life cycle of conifers can be described as the alternation of a haploid gametophyte generation with a diploid sporophyte generation. Two types of cones are produced, male or pollen cones and female or seed cones. In the pollen cones, special cells undergo reduction division resulting in a great number of haploid spores that are shed as pollen grains. Each of these is a male gametophyte. All conifers are pollinated by wind, and in a coniferous forest, pollen grains are shed in such numbers that they can be seen as yellow clouds during peak periods of dispersal. A short time after pollen is dispersed, the male cones shrivel and fall from the tree.

The female cone is made up of flat scaly divisions attached to a central column. On the upper surface of each scale, two mounds of tissue develop and in each a single cell undergoes reduction division resulting in four haploid spores. Three of these disintegrate, and the remaining cell grows into the female gametophyte bearing one or more egg cells. The pollen grain or male gametophyte, having been transported to the female cone by wind, develops a pollen tube that carries the sperm to the egg cell. The sperm and egg unite to form a zygote that grows into the young sporophyte. The embryonic sporophyte is surrounded by a layer of nutritive tissue and a hard outer covering. This is the seed that may become one of next year's crop of conifers. Pollination of most conifers usually occurs in May. In the temperate zone, the seeds of most are shed in the autumn following pollination, but in pines they are shed in autumn the year after pollination. Each scale of the female cone bears two seeds with flat winglike projections that aid in dispersal by wind.

Flowering Plants

The flowering plants are referred to as angiosperms, a term that means "seeds in a receptacle." This alludes to the fact that their seeds develop inside the ovary, which matures to become the fruit. The flowering plants are the most recently evolved of all the plant groups—their oldest fossils are about 130 million years old. Their ancestors were ancient gymnosperms that are today extinct. The time when dinosaurs roamed the earth was the age of gymnosperms, but today we are in the age of angiosperms. They dominate world vegetation with the greatest number of individuals and the greatest number of species. Unlike the gymnosperms, they grow in a variety of forms including trees, shrubs, vines, herbs, and non-green parasites.

Flowering plants range in size from eucalyptus trees at over 300 feet high (90 m) and 60 feet (18 m) around at their base, to floating duckweeds that are $1/_{25}$ of an inch (1 mm) in diameter. They have covered the earth and have adapted to an amazing range of environments from the arctic to the tropics. Some are water plants that grow completely submerged or in waterlogged soil. Others are drought-resistant plants that can survive the arid deserts of the world. Some species called air plants grow on the trunks of tropical trees with roots that hang freely and absorb water vapor from the air. Flowering plants dominate most of the terrestrial habitats on earth today.

Angiosperms are true land dwellers as are the mammals of the animal kingdom. In ferns, water is required for the sperm to swim to the egg in order to complete the life cycle. They resemble the amphibians of the animal world in this regard. In angiosperms, the link with aquatic ancestors has been severed. The male gametophyte with sperm cells is delivered to the egg cell by wind, insects, or some other animal pollinator; no water is necessary. Mammals, birds, and insects have evolved in close association with flowering plants. The rise of herbivorous mammals was dependent on the development of grasses and other herbs. Birds have evolved with their main sources of food, the seeds and fruits of flowering plants or the insects that feed on leaves and fruits. Insects and angiosperms have a remarkable history of codependence and coevolution. Humans' domestication and subsequent cultivation of certain flowering plants was the initial step in the development of modern civilization.

The flower is the organ of reproduction for the angiosperms. A typical flower consists of an outer ring of green leaflike parts called the sepals. Collectively the sepals make up the calyx. Its function in the bud is the protection of the delicate inner parts. Inside the calyx is the corolla, which is made up of individual parts called petals. The corolla in many flowers is brightly colored and associated with sweet-smelling nectar glands—features that attract insects and other pollinators. Inside the corolla is a ring of stamens, the male parts of the flower, each of which consists of a slender stalk, the filament, which supports the anther. In the center of the flower is the female

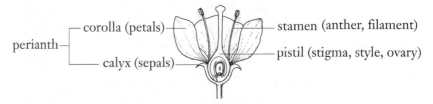

2.16. Angiosperm flower

reproductive structure, the pistil (there may be more than one), with an enlarged lower part, the ovary, an elongated neck, the style, topped by a sticky surface, the stigma, which receives the pollen grains. One or more ovules develop inside the ovary, and these will eventually become seeds (fig. 2.16).

As in other plant groups, angiosperms have alternation of generations as part of their life cycles. The anther contains four pollen sacs or sporangia that produce haploid male spores. These are shed as pollen grains that become the male gametophytes. Inside each ovule there is a haploid female gametophyte consisting of only eight cells, one of which is the egg cell. During pollination, the pollen grain is transported to the stigma where it germinates. A pollen tube containing two sperm cells grows through the style to the egg cell. One of the sperm cells unites with the egg to form a diploid zygote that will become the next sporophyte generation. The other sperm cell unites with two other haploid cells, and the resulting triploid cell grows into a food-storage tissue known as the endosperm. At this point the ovule becomes a seed.

The stored food in the endosperm helps assure survival of the sporophyte seedling. Humans also have found this tissue to be an important source of food. It provides the nourishment of the cereal grains, which include wheat, corn, rice, oats, and rye. Coconut meat is endosperm tissue, and coconut milk is endosperm that did not form into cells. The food value of peas, beans, peanuts, and other legumes is in endosperm that has been transformed and redeposited in two very thick seed leaves known as cotyledons.

Monocots and Dicots

There are two basic groups of flowering plants, the monocotyledons (monocots) and the dicotyledons (dicots), that are fairly easy to recognize in the field. Since there are twice as many dicots as monocots, the plants most often seen are dicots. Cotyledons are seed leaves, and, as the names suggest, monocots have one and dicots have two. A seed usually consists of a seed coat, food-storage tissue, and the embryonic sporophyte with its first leaves. Sometimes all the stored food is converted into cotyledons as in beans and peas. The time to observe cotyledons is shortly after the seed germinates; the first structures to appear above the ground are the cotyledons, and whether there is one or two will be obvious.

There are features other than seed leaves that can be used in the field to distinguish between these two groups. The leaves of monocots have veins that are parallel from the base to the tip of the leaf; in dicot leaves the veins are branched into a network. The most reliable and easiest way to distinguish between these two groups is by the number of flower parts. In monocots the flower parts are in multiples of three. A typical species could have three sepals, three petals, and six stamens (fig. 2.17). Among the spring flowers, wild trilliums are an example of this type. In some species the sepals are modified so as to be indistinguishable from the petals; the lily flower appears to have no sepals but rather six petals and six stamens. The monocotyledons include such familiar plants as trilliums, irises, spiderworts, and orchids. The orchids may be the second largest family of flowering plants and the most highly evolved of the monocots.

The flower parts of dicotyledons are in multiples of four or five (fig. 2.18). Almost all of the trees and shrubs and most of the herbs

2.17. Monocotyledon flower (lily)

2.18. Dicotyledon flower (buttercup)

are dicots. A common flower type of these plants could have five sepals, five petals, and five to ten stamens. The aster family is the largest family of flowering plants and the most highly evolved of the dicotyledons. The flowers of this family are very small and so tightly clustered that each cluster gives the appearance of a single flower. For example, the big-leaved aster *(Aster macrophyllus,* fig. 6.1) looks like a flower with lavender petals and a yellow center. Actually each of the "petals" is a flower, and the yellow center is made up of many tiny flowers, each consisting of a five-lobed corolla, five stamens, and a pistil with a curling two-parted stigma.

The number, shape, color, and location of flower parts are the chief traits used in the identification of herbaceous plants. Leaf characteristics are the most important features used in the identification of trees and shrubs. These will be discussed further in chapters 5 and 6.

3

Adaptations for Survival

Genetic Variability

Asexual Reproduction

Most species of plants produce offspring by both asexual and sexual means. In asexual reproduction, there is no union of male and female sex cells. Consequently the genetic makeup of the offspring is exactly the same as that of the parents. It can be as simple as a piece of the parent plant breaking off and growing into a new plant. This type of asexual regeneration is exhibited when some species of trees, such as aspens *(Populus spp.)* and willows *(Salix spp.)*, are cut and sprouts arise from the stumps.

In other species, asexual reproduction is a more structured part of the life cycle. For example, walking fern *(Asplenium rhizophyllum)* grows mostly on rocks in the eastern deciduous forest (fig. 3.1). It is called walking fern because at every point where the tip of a frond touches down, roots form and a new plant grows. In the houseplant known as maternity plant *(Kalanchoe diagremontiana)*, small plants develop along the margins of leaves. Each of these drops to the ground, produces roots, and grows into a clone of the parent.

In the most complex

3.1. Walking Fern *(Asplenium rhizophyllum)*

form of asexual reproduction, seeds are produced without the union of male and female sex cells. Seeds are normally the result of sexual reproduction, but in some species, as white trillium *(Trillium grandiflorum,* fig. 6.19), false Solomon's seal *(Smilacina racemosa,* fig. 7.22), and hawthorn *(Crataegus spp.),* an ordinary cell in the ovary begins to divide and form an embryo that will eventually become a seed. This type of nonsexual reproduction has the advantage of seed dispersal mechanisms for spreading into new areas. The seeds will germinate and, like other forms of vegetative reproduction, grow into a plant that is an exact genetic duplicate of the parent plant.

Maintaining an exact genetic composition has advantages in some situations. In harsh environments where there may be a close coordination between plant features and the demands of the habitat, a slight variation in plant form may result in extinction. Also, in rigorous arctic environments where there may be a scarcity of insect pollinators, asexual reproduction is sometimes the only road to survival. However, plant species that reproduce by asexual means only are evolutionary dead ends. In the absence of sexual reproduction, they cannot present a variety of genetic combinations for selection by a changing environment. Since they have only one combination, the only response they can make to climatic change is extinction.

Sexual Reproduction

In sexually reproducing species, there is a reshuffling of genes with each generation, providing a variety of genetic types. As a result, when there is a substantial change in the environment, although some members of the species will die, others may have genetic combinations enabling survival in the changed conditions. This advantage led to the origin of sexual reproduction and to its persistence in most living things today. All groups of plants, including the algae, mosses and liverworts, ferns and fern allies, coniferous plants, and flowering plants, have well-developed sexual systems. Since the flowering plants are the most commonly seen plants in the forest, their sexual reproduction will be discussed in more detail.

Methods of Cross-Pollination

Plant species with the greatest numbers of genetic variations are best adapted for long-term survival. In flowering plants, the key to maintaining the greatest number of variations is cross-pollination. The transfer of pollen, which carries the male sex cells to the pistil, where the female sex cells are located, is pollination. Most flowering plants have both male and female parts in the same flower. Cross-pollination occurs when the pollen from one plant is transferred to the pistil of a different plant. When pollination is from the same plant, there is less variation among offspring than when the pollen is from another plant. Numerous growth habits have evolved that promote cross-pollination.

Animal Pollinators

Insects

The most important agents of cross-pollination are insects. They have been associated with flowering plants for at least forty to sixty million years. During this time, many specialized relationships have evolved in which plants and insects are mutually dependent on one another for survival. For example, the female yucca moth is attracted to the creamy-white, sweet-smelling flowers of the night-blooming yucca plant. She collects a ball of pollen from the flower of one plant, then flies to another. At the second plant she places the ball of pollen on the stigma and pierces the ovary wall to lay her eggs inside. The developing moth larvae feed on the tissue of the ovary, consuming about 20 percent of the young seeds. This does not endanger the plant and is a modest price to pay for the advantage of cross-pollination.

Most of the time, cross-pollination by insects is not as deliberate as the above example. Visits to flowers are usually for nectar or pollen as sources of food. Insect-pollinated plants have sticky pollen that adheres to the body of the pollinator. When the insect visits another flower of the same species, it accidentally brushes against the stigma and the pollen is transferred.

Bee pollination. Bees are the most important of the insect pollinators. There are at least twenty thousand species, all of which must visit flowers for food. Plant species pollinated by bees have evolved special types of flowers that are easy for bees to find and on which they can land. These insects cannot recognize the color red, but they can see ultraviolet light, which is invisible to the human eye. Flowers that have developed in response to bee pollinators are usually yellow or blue. Some have special markings visible only under ultraviolet light that highlight the location of nectaries. These

3.2. Wild Geranium *(Geranium maculatum)*

are glands that produce a sweet substance called nectar that many species of insects use as food. Other bee flowers, like wild geranium *(Geranium maculatum,* fig. 3.2), have large lobes that serve as landing platforms, and special markings called nectar guides that are like road signs directing bees to the nectar glands.

Most members of the orchid family (Orchidaceae) are insect-pollinated, and evolution has made them masters of deception. The flowers of some species have odors that lure the hapless bees to them, and others have clusters of yellow hairs that bees mistake for pollen. Other families of plants commonly offer nectar or pollen or both to pollinators, but many members of the orchid family provide neither. A bizarre deception is seen in the looking-glass orchid *(Ophrys speculum)* of southern Europe and Algeria. The flowers of this orchid resemble the females of a species of wasp *(Scolia ciliata)*. Not only do the flowers look like the female wasp, they also emit a similar odor. The male wasps emerge early in the season before the females. They are attracted to the flowers and attempt to copulate with them. Then they repeat the process with a second flower where they deposit pollen from the first.

Moth and butterfly pollination. Moths and butterflies are very important pollinators. Most moths are night-flying creatures that have coevolved (developed in response to one another) with night-blooming plants. Since bright colors are not visible at night, most moth-pollinated flowers are white or of a pale color that will stand out against a dark background. Moths have a well-developed sense of smell, and flowers that attract them emit powerful fragrances only after sunset. Both moths and butterflies have long sucking tongues that permit them to reach the nectar in narrow tubular flowers. Moth-pollinated plants are most common in tropical forests.

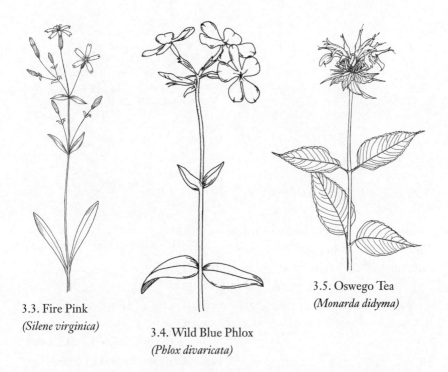

3.3. Fire Pink
(Silene virginica)

3.4. Wild Blue Phlox
(Phlox divaricata)

3.5. Oswego Tea
(Monarda didyma)

Flowers pollinated by butterflies are usually showy, fragrant, and day-blooming as are the flowers pollinated by bees. Unlike bees, some butterflies can see the color red and visit red and orange flowers as well as blue and yellow ones. Woodland plants with colorful flowers often pollinated by butterflies are fire pink *(Silene virginica,* fig. 3.3), wild blue phlox *(Phlox divaricata,* fig. 3.4), and Oswego tea *(Monarda didyma,* fig. 3.5). One group of moths, the hawkmoths, are active during the day and may visit the same flowers as butterflies.

Fly and beetle pollination. The food of beetles and flies is frequently decaying fruit, dung, and dead animals. Plants that have been influenced by these insects in their evolution often have flowers with the unpleasant odors of rotting tissue. The sense of smell in beetles is more highly developed than the sense of sight, and the flowers they visit are usually not brightly colored. Most beetles do not have mouth parts suited for obtaining nectar, especially from tubular flowers, so they feed on flower parts or pollen. There are at least thirty thousand species of plants pollinated by beetles,

3.6. Tulip Tree *(Liriodendron tulipifera)*

with more being discovered each year. Some plants pollinated by beetles are the magnolias *(Magnolia spp.)* and tulip tree *(Liriodendron tulipifera,* fig. 3.6).

3.7. Dutchman's Pipe *(Aristolochia macrophylla)*

They also can be seen frequently on the flowers of common elderberry *(Sambucus canadensis)* and wild roses *(Rosa spp.)*.

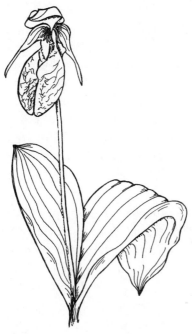

3.8. Moccasin Flower *(Cypripedium acaule)*

A variety of flower types are pollinated by flies. Like the beetles, they have a highly developed sense of smell. Two plants of the same family that are commonly pollinated by flies are skunk cabbage *(Symplocarpus foetidus)* and jack-in-the-pulpit *(Arisaema triphyllum,* fig. 6.4). The flowering structure in each of these consists of many male and female flowers on a central column enclosed in a sheath. An odor similar to rotten meat attracts flies to these plants. In skunk cabbage, the central column generates heat that serves to intensify the odor. Other fly-pollinated species, like Dutchman's pipe *(Aristolochia macrophylla,* fig. 3.7) and moccasin flower *(Cypripedium acaule,* fig. 3.8), are trap flowers that have odors that lure flies into a chamber of the flower. In order to enter, the flies must come into contact with the stigma, and they are dusted with pollen as they escape to visit the next flower.

Birds

In different parts of the world, many species of birds are specialized to feed on flower parts, flower-eating insects, or nectar. As with most insect pollinators, cross-pollination by birds is accidental. In North America the most important bird pollinators are the hummingbirds, whose most important source of food is nectar. They have long slender beaks that can penetrate to the base of the longest tubular flowers. Hummingbirds have a well-developed sense of color and can see the reds, but they have a very poor sense of smell. Consequently, plants pollinated by hummingbirds often have red flowers with little or no odor.

The red color of hummingbird flowers serves a dual purpose: it is a highly visible welcome mat for the birds and it discourages insects because they cannot see red colors. Hummingbirds are heavier and require more energy to fly than insect visitors, so the flowers they pollinate must produce large quantities of nectar. It is usually found in long tubular flowers or spurs that cannot be reached by insects. Even if insects could reach the nectar, they would be ineffective cross-pollinators of hummingbird flowers because the quantity of nectar would satisfy their need and they would not visit other flowers. Some typical hummingbird-pollinated flowers are wild columbine (*Aquilegia canadensis*, fig. 3.9), passion flower (*Passiflora incarnata*), and bird of paradise (*Strelitzia reginae*).

Genetic Safeguards

Although cross-pollination by insects or birds is usually very dependable, there is still the possibility that pollen

3.9. Wild Columbine (*Aquilegia canadensis*)

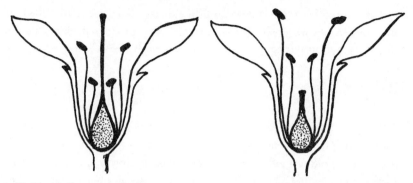

3.10. Type 1 flower; type 2 flower

from the anther could reach the stigma in the same flower. This is called self-pollination, and many plant species have physiological or structural features to keep it from happening. One way has been through the development of different genetic strains within a species. The pollen of one strain is physiologically rejected by the stigma of any flower on the same plant. It must reach the stigma of another plant before it can grow a pollen tube that will result in the production of a viable seed. This is called self-incompatibility and is common among species of wild plants.

A strategy to avoid self-pollination in some insect-pollinated plants is the growth of pistils and stamens with different lengths. In bluets (*Hedotis caerulea*), there are two types of flowers in approximately equal numbers: those with long-styled pistils and short stamens, and those with short-styled pistils and long stamens (fig. 3.10). This greatly reduces the probability of pollen reaching the stigma in the same flower. However, as added insurance, this condition is usually accompanied by self-incompatibility. Fertile seeds can be produced only when pollen from a type 1 flower reaches the stigma of a type 2 flower. In some species there are three types of flowers with regard to length of pistils and stamens.

Separation of the Sexes

Some plants have stamens (male parts) and pistils (female parts) on separate flowers. These are called monoecious plants. Sometimes the male and female flowers are in different locations on the same plant as in the oaks (*Quercus spp.*), beech (*Fagus grandifolia*), and the birches (*Betula spp.*). Although self-pollination is less likely in these species, unless they are self-incompatible, it could occur.

In other species, the individual plants bear either male or female flowers but not both. These are called dioecious plants, and this arrangement is a guarantee that self-pollination can never occur. The disadvantage is that only half of the population can produce seeds since male and female plants are present in about equal numbers. An herbaceous woodland species with unisexual flowers is early meadow rue *(Thalictrum dioicum,* fig. 3.11). Among woody dioecious plants are some of the willows, aspens, and the hollies *(Ilex spp.).* American holly *(Ilex opaca)* with its spiny leaves has become a symbol for the Christmas season in North Amer-

3.11. Early Meadow Rue
(Thalictrum dioicum)

ica. Its bright red berries on the female plants are a characteristic feature that may not be present unless there is a male tree in the vicinity. Plants with unisexual flowers, both monoecious and dioecious, are more common among wind-pollinated species.

Wind Pollination

Wind is an extremely inefficient agent of pollination. Whether or not an individual pollen grain reaches a stigma is purely a matter of chance. To increase poor odds, wind-pollinated plants produce great quantities of pollen. So much is produced that even the most remote stigma is likely to be dusted. Only in this way can a seed crop big enough to sustain the species be assured. By one estimate, a single male flower of an oak tree contains over forty thousand pollen grains. With thousands of male flowers on a tree, it releases millions of grains each spring. In a study of hazelnut shrubs, it was calculated that a single male flower yields about two million pollen grains with a total of 600 million for an average-sized shrub.

Since their evolution was not influenced by animal pollinators, wind-pollinated flowers do not have colorful petals, do not produce nectar, and do not have a fragrance. The stamens are usually long with anthers freely exposed to the air, as in the box elder (*Acer negundo*). In wind-pollinated trees, the male flowers are often in long dangling clusters called catkins. These sway in the slightest breeze, releasing their pollen. Catkins can be seen in the oaks, birches, aspens, and beech. The stigmas of wind-pollinated flowers are usually extensively branched exposing the maximum amount of surface. This increases their efficiency in trapping pollen grains. The ovary usually has only one ovule so only one pollen grain is necessary for successful pollination.

In tropical rain forests, wind pollination is noticeable by its absence. At least two factors make these areas unsuitable for wind pollination: the plants do not have a leafless season and the almost daily rain showers would completely wash the pollen out of the air. In contrast, temperate deciduous forests, where most of the trees are wind pollinated, have a distinct leafless season. These trees usually release their pollen in spring before there are leaves to interfere with its dispersal. In addition, rain showers are less frequent to wash the pollen out of the air. Another factor that may be less noticeable to an untrained observer is that trees of the same species are closer to one another in the deciduous forest than in the tropical forest.

Seed Dispersal

A discussion of seed dispersal cannot be complete without a clear understanding of the relationship between fruits and seeds. Seeds develop within the ovaries of flowering plants, and the ripened ovary is a fruit. There are two types of ovaries, fleshy and dry. Fleshy fruits have thick walls that are sometimes colorful when the seeds are mature. Some of these are sweet and juicy and commercially identified as fruits at the grocery store. Others, like tomatoes, cucumbers, and green peppers, are commonly called vegetables. Dry fruits are those in which the ovary wall is usually dry when the seeds are mature. In cultivated plants, beans and peas are of this type, as is black locust (*Robinia pseudoacacia*, fig. 7.8) among forest trees. In some dry fruits, the ovary contains only one seed and at maturity the ovary wall becomes part of the seed coat. Maple and ash trees have seeds that botanists classify as fruits.

Most seeds fail to complete their evolutionary mission: the growth of a new plant. Consider, for instance, an oak tree that may produce thousands of acorns each year. Suppose only ten of these grew into reproducing adult trees. In the next generation, presume further that each of the ten yielded ten reproducing offspring. If this process continued for ten generations, the resulting number of oak trees would be more than the entire surface of the earth could support. The numbers of seeds produced by some species of herbaceous plants are even more impressive.

In spite of this high rate of failure, seeds continue to be the most important means of reproduction and dispersal among flowering plants. They provide the plant with mobility, allowing the species to colonize new areas and increase its range of distribution. Still another advantage of seed dispersal is that seedlings escape from competition with the parent plant. In response to these advantages, and perhaps others, a variety of seed dispersal mechanisms have arisen in plants. The two most common agents of dispersal are animals and wind.

Dispersal by Animals

Animals are the most effective agents of seed dispersal. There are at least two reasons for this. First, migrating birds and mammals move at predictable seasonally regulated intervals. Over a long period of time, this could result in the evolution of plants with seeds that are mature at the time of migration. Second, since mammals usually move from one favorable environment to another, the seeds they transport are likely to be deposited in an area that is favorable for germination and growth. In some deciduous forests, the seeds of as many as 60 percent of the trees and nearly 70 percent of the herbaceous plants are dispersed by animals. Dispersal is by three methods: (1) ingestion, (2) adherence to the outer surface of fur, feathers, or feet, and (3) transportation and storage as a food reserve.

Ingestion. Fleshy fruits have evolved chiefly as organs for seed dispersal. Animals are attracted to them as food sources, and then seeds are transported in the intestines of the animals. Many seeds pass through animal digestive tracts unharmed. By one estimation, fruit eaters are responsible for seed dispersal in 12.5 percent of the flowering plants in northeastern North America. Birds are the most important of these, but mammals and reptiles are also fruit eaters. In over 70 percent of the plants that have bird-disseminated seeds, fruit ripening coincides with the onset of fall bird migration. A native species that has seeds dispersed by birds is wild black cherry *(Prunus serotina,* fig. 7.10). It has been introduced into Europe where its wide dispersal by birds has made it a nuisance species in some areas.

Sometimes seeds pass through the digestive tracts of browsing or grazing animals. Such passage, whether in grazers or fruit eaters, often improves germination by softening hard seed coats.

Adhesion. Seeds dispersed in this manner are usually from herbaceous plants and have hooks, spines, or a sticky surface. This type of dispersal is typical of plants on the forest floor like enchanter's nightshade *(Circaea lutetiana)* and tick-trefoil *(Desmodium canadense).* The seeds of these species readily become attached for a free ride on the fur of any passing animal or the clothing of any field naturalist (fig. 3.12).

Food storage. Some seeds are dispersed by animals that transport and store them as a reserve food supply. Most of the store is eaten, but the animal usually gathers more than is used or forgets where some of it is stored. These give rise to the next generation of plants. Large seeds such as acorns, beechnuts, hickory nuts, and walnuts are collected by squirrels and chipmunks. The distance they are carried beyond the parent tree is usually not more than a few yards. Jays have been known to carry acorns for a distance of $2^1/_2$ miles (4 km) or more.

The seeds of many herbaceous plants have energy-rich nodules that attract ants. The ants carry the seeds to their nests where they consume the nodules and leave the seeds unharmed. Thus they not only disperse the seeds, but, like the chipmunks and squirrels, they plant them as well. The distance seeds are carried by ants is not great, but it is usually well beyond the range of competition with the parent plant.

Seed dispersal by ants is much more common than was once believed. In some deciduous forests of New York and West Virginia, they are responsible for dispersal of 36 percent and 30 percent, respectively, of the herbaceous plants. Common woodland plants

3.12. Seed of Enchanter's Nightshade *(Circaea lutetiana),* Tick-trefoil *(Desmodium canadense)*

with seeds dispersed by ants are the wild trilliums *(Trillium spp.)*, violets *(Viola spp.)*, wild ginger *(Asarum canadense,* fig. 6.16), bloodroot *(Sanguinaria canadensis,* fig. 6.17), and twinleaf *(Jeffersonia diphylla,* fig. 3.13).

Dispersal by Wind

Wind is the second most important agent of seed dispersal. It is less efficient than animals for two reasons: (1) wind is highly variable and unpredictable, and it may not be present at the best time for dispersal, and (2) wind dispersal is random, with many seeds falling in areas unsuitable for germination. Despite these shortcomings, species with seed modifications for wind dispersal are common. A few of these modifications are described in the following paragraphs.

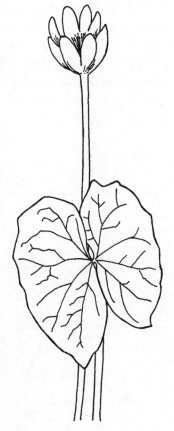

3.13. Twinleaf *(Jeffersonia diphylla)*

Size. Very small seeds can be seen as an adaptation for dispersal by wind. The smaller the size, the greater the ease of dispersal. The orchid family has the smallest known seeds. A woodland species, rattlesnake plantain *(Goodyera repens)*, has seeds that weigh 0.000002 grams each. It would take 500,000 seeds to weigh one gram and 14,187,500 to weigh one ounce (28.4 g). These dustlike seeds can be widely dispersed in the slightest breeze.

Parachutes. A common adaptation for dispersal by wind is a tuft of hairs that function as a parachute. This type of dispersion is found in trees like the aspens and willows, and woodland wildflowers such as big-leaved aster *(Aster macrophyllus,* fig. 6.1) and blue-stem goldenrod *(Solidago caesia,* fig. 6.25). About 16 percent of North American plants are dispersed by seeds with parachutes.

Wings. Thin membranous wings are effective mechanisms for dispersal by wind. These slow the rate at which the seed falls, giving the wind time to carry it sometimes for long distances. The firs *(Abies spp.)* and spruces *(Picea spp.)* have winged seeds. The one-seeded fruits of the maples *(Acer spp.)*, ashes *(Fraxinus spp.)*, elms *(Ulmus spp.)*, birches, and tulip tree have wings. These whirl like tiny propellers and in strong winds may be carried for several miles. In the deciduous forest, the seeds of 35 percent or more of the trees are dispersed by wind.

Explosive Fruits

Probably the least effective method of seed dispersal, but a highly interesting one, is the explosive seedpod that flings seeds in all directions. This type of dispersal is exhibited by orange touch-me-not *(Impatiens capensis)*. It grows abundantly in eastern North America along the wet margins of forests and in roadside ditches. At maturity, the seed capsule of this plant may be $1^1/_4$ inches (3 cm) long, and at the slightest touch, as it sways in the wind, it bursts explosively (fig. 3.14). The capsule splits from the base into five segments that roll inward forcefully hurling the seeds for a distance of 8 to 10 feet (2.4–3 m).

Witch-hazel *(Hamamelis virginiana)* and some species of violets also eject their seeds forcefully but not by an explosive seedpod.

3.14. Touch-me-not *(Impatiens capensis)*

4

Soil, Climate, and Vegetation

Soil Formation

Soil has been called the umbilical cord of life. It is the transition zone between the living and nonliving worlds. It is a zone of both an ending and a beginning. An essential component of soil, organic matter, consists of the dead bodies and body parts of plants and animals. As they decompose, the basic substances that originally gave them life are returned to the soil. These are absorbed by the roots of plants and, through the process of photosynthesis and other physiological processes, are converted into the food that gives life to all living things on earth.

Soil is initially formed by the breakdown or weathering of rock, but it is much more than simply weathered rock. It is a medium that will support plant roots, and it must provide most of the physical substances needed by plants. In order to perform this function, soil must contain components other than weathered rock. Organic matter, as it decomposes, becomes a finely divided, black substance called humus. In its final stages of decomposition, mineral nutrients are released. A third essential part of the soil consists of living organisms. The agents responsible for reducing organic matter to humus are the bacteria and fungi of decay. Every square foot of soil contains millions of these organisms. Since they must have oxygen and water, these substances also are essential parts of the soil. All of these ingredients are interrelated in a way that makes the soil a very complex ecosystem.

The weathering of rock to soil-sized particles is the result of many factors, and it requires a very long period of time. Physical changes in the environment such as heating and cooling, wetting and drying, freezing and thawing, and erosion by wind and water all contribute to the process. Chemical changes such as oxidation, reactions with water, and the action of acids formed by the union of carbon dioxide and water also contribute to soil formation. In addition, rock is broken down by living things. For example, plant roots may grow into cracks, then increase the sizes of the cracks as they expand with growth. Most areas of the earth today are covered with a layer of soil. Notable exceptions are steep, high mountain peaks and regions where glaciers have scraped away all loose material exposing the underlying bedrock.

The material from which soil develops, the parent material, may not be the bedrock but is very often material that has been transported from one place to another by wind, water, or ice. In southern Canada and the northern United States, the soil has developed mainly from material brought in by glaciers during the past million years. In river valleys and deltas, soil has developed from material washed down from higher elevations by water. In parts of midwestern North America, especially in the corn belt, soil has developed from deep layers of wind deposits.

Soil Texture

In all types of parent material, the weathering process results in soil particles of many different sizes. The proportion of different sized particles is soil texture, a property of great importance. The U.S. Department of Agriculture Soil Conservation Service has classified soil particle size, from smallest to largest, as clay, silt, sand, and gravel. Individual clay particles are so small that they can be seen only with an electron microscope. The largest of them would require 12,500 particles placed side by side to equal an inch (25 mm). Silt particles can be seen with a regular microscope, and 500 of the largest ones would equal an inch (2.5 cm). Particles larger than silt and ranging to $^2/_{25}$ of an inch (2 mm) in diameter are classified as sand or gravel. These are visible to the naked eye. Agriculturally, soils with high proportions of clay are difficult to turn over with a plow and are referred to as "heavy" soils, while those with high concentrations of sand are easily turned and are called "light soils."

Soil texture influences several soil properties that are essential for plant growth. Most of the water used by plants is held as a film around soil particles. The film is about the same thickness regardless of the size of the parti-cle. A fine-textured soil with a high proportion of clay can hold more water than a coarse-textured one with a high proportion of sand. Thus, in time of drought, plants growing in a sandy soil will begin to wilt before those grow-ing on soil with a high clay content.

Soil texture also influences the amount of air in the soil. In a fine-textured soil, the air spaces are smaller and the movement of both water and air are restricted. This may inhibit growth of both plant roots and microor-ganisms. In addition, plant roots must continuously grow into new areas in order to obtain water. A tightly packed, fine-textured clay soil can retard penetration of plant roots.

The soils in which plants grow best have characteristics of both sand and clay. These include the large air spaces and easy root penetration of sandy soils and the water-holding capacity of soils with a high clay content. These features occur naturally when sand, silt, and clay are in the appropri-ate proportions. This type of soil is called loam, and it is the best type for farming. An aspect of loam is that the individual soil particles stick together in small clumps of various sizes. If a handful is examined, it appears to be made up of small lumps. Each of these behaves somewhat like an individual soil particle, and it is this structure that gives the soil its desirable qualities for plant growth.

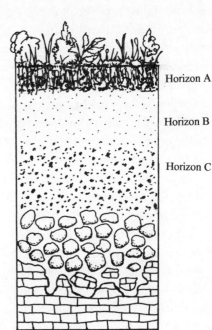

Horizon A

Horizon B

Horizon C

Soil Profile

Parent material that has been ex-posed to weathering for a long pe-riod of time in temperate climatic conditions will develop a series of layers or horizons that collec-tively are called the soil profile. These have been identified as the A, B, and C horizons (fig. 4.1). The A horizon is the topsoil. It is the layer that has most of the or-

4.1. Soil profile

ganic matter or humus and the greatest number of soil organisms. The roots of plants, especially crop plants, absorb most of their water and mineral nutrients from this layer. Water from rain and melting snow percolates through the topsoil dissolving the soluble minerals and carrying them downward. For this reason, the A horizon is sometimes called the zone of leaching. Most of the food humans must have for survival depends on the topsoil.

Below the topsoil is the B horizon or subsoil. Because the minerals and fine clay particles leached from the topsoil are deposited here, the B horizon is sometimes called the zone of deposition. Although some organic matter may be carried downward to this layer by percolating water, the amount is very small. This is not a fertile medium for plant growth. If sufficient iron is present, minerals leached from the topsoil may be cemented together to form an almost concrete-like hard pan in the lower levels of the subsoil. When the topsoil is removed by strip mining, construction, or other human activities, the exposed subsoil often becomes barren and eroded.

Beneath the subsoil is the C horizon or parent material. This may be bedrock or, more likely, material that has been transported to its present location. In any event, all parent material, at some point in time, had its beginning as rock. The nature of the soil will obviously be influenced by the chemical composition of the rock from which it was formed. Soils developed on limestone, or other rock with a high calcium content, are alkaline in reaction because calcium neutralizes acid compounds.

The soils of much of northeastern North America originated from granite or other igneous rocks that do not contain calcium. Consequently, this area is much more subject to damage from acid rain than are regions where the soil developed from calcium-containing rocks. In the granitic Adirondack Mountains, acidic water draining into lakes has eliminated the fish and amphibian life from many of them.

The soil profile can sometimes be observed in road cuts or where there have been excavations. The A horizon will appear darker in color because of its humus content. In the eastern United States, the topsoil may be 3 to 14 inches (7.5–35 cm) thick, depending on the amount of erosion that has taken place. When Europeans arrived on this continent, the topsoil over most of North America was considerably thicker than it is today. Below the topsoil, the subsoil will appear lighter in color. As a result of the leaching of fine clay particles from above, the subsoil has a high clay content and may be very compact. The lower level of the subsoil grades into weathered bedrock or other parent material.

Climate and Vegetation

The type of soil profile that develops, regardless of the type of parent material, depends in large measure on climatic conditions. Thus, in a region subjected to a given set of climatic variations, there will be similar soil types and similar vegetation throughout. In North America there are several such geographic regions called biomes. They include the arctic tundra, northern evergreen forest or taiga, eastern deciduous forest, southern evergreen forest, grassland, and desert.

Arctic Tundra

One of the characteristic features of the tundra is the presence of a permanently frozen or permafrost layer 1 to 2 feet (30–60 cm) below the surface. Above this zone, the soil thaws during a growing season that is often less than two months in length. Plants are low growing, usually no more than 1 foot (30 cm) high. They are mostly herbaceous perennials with brightly colored flowers that develop in the twenty-four hour daylight period of the growing season. Because of the low annual temperatures, decomposition is very slow. Consequently the top layer of soil contains a high proportion of undecomposed organic matter. The mixing caused by seasonal freezing and thawing produces a soil that typically has no recognizable profile.

Northern Evergreen Forest

South of the tundra and, like the tundra, forming a band around the top of the earth, is the northern evergreen forest. The Russians call this biome the taiga, and many American ecologists have begun using that term. It is also sometimes called the boreal forest. In North America it extends from Alaska to Newfoundland southward to northern New England, northern Michigan, and northern Minnesota. The climate in most of the boreal forest is severe, with cold winters and short summers. The most widespread tree is white spruce *(Picea glauca)*. In the southeastern part of the forest, spruce grows with balsam fir *(Abies balsamea)*. The most common pine is jack pine *(Pinus banksiana)*, and the main broad-leaved trees are paper birch *(Betula papyrifera)*, balsam poplar *(Populus balsamifera)*, and quaking aspen *(Populus tremuloides)*.

The evergreen trees of the taiga contain resinous compounds that are highly flammable. These are present in both the wood and the leaves so that the whole tree will burn. Fires started by lightning were common long before humans arrived in North America. As a fire flashes from tree to tree in what is called a crown fire, whole sections of the forest may be reduced to charred stumps and ashes. After North America was settled by Europeans, fires from logging operations were more frequent than those caused by lightning. In the early days of lumbering, several towns and cities were threatened or completely burned by uncontrolled crown fires (see chapter 1).

When a section of the mature forest is destroyed, especially by fire, one of the first herbaceous plants to appear is fireweed *(Epilobium angustifolium)*. The pink to purple flowers soon set the scorched landscape ablaze with a different kind of color (fig. 4.2). Fireweed and other herbaceous plants are accompanied by seedlings of paper birch, quaking aspen (fig. 4.3), and balsam poplar. After a few years, autumnal col-

4.2. Fireweed *(Epilobium angustifolium)*

4.3. Quaking Aspen *(Populus tremuloides)*

oration of their leaves outlines the burn scar in bright yellow contrast to the green of the unburned area. Seeds of spruce and fir germinate, and seedlings grow among the birch and aspens. These species are slow growing, but they will eventually shade out the broad-leaved trees. In about three hundred years, a mature spruce-fir forest will reclaim the burned area.

The forest floor usually has a carpet of needles, twigs, and cones that decompose very slowly because of the low average annual temperature. When rainwater percolates through this material, it becomes very acid. Much of the iron, aluminum, and other minerals in the topsoil are leached away leaving mostly sand. The leached minerals are deposited in the subsoil, giving it a tan or brownish color. Soils formed in this way are called podzols and the process is called podzolization. Podzol soils are a characteristic feature of the taiga. The evergreen trees are adapted to survive in these soils, but when the forest is cleared the soil is so poor that it will not support cultivated crops.

Eastern Deciduous Forest

Southeast of the boreal forest is the eastern deciduous forest. Several million years ago, the deciduous or summergreen forest was worldwide in distribution in the north temperate zone. As a result of glaciation and other climatic changes, it is today reduced to three segments; western Europe, central Asia, and eastern North America. Currently, the most extensive and best preserved of these are the forests of eastern North America. As in the taiga, the soil-forming process is podzolization. The chief difference is that in the deciduous forest the rate of decomposition is greater and the leaves contain more calcium. The result is that percolating rainwater becomes less acid and leaching from the topsoil is less severe. The soils typically develop profiles with A, B, and C horizons as described earlier. As early settlers discovered, when deciduous forests are cleared, the soil can be excellent for cultivated crops. See chapters 5 and 6 for more information on the summergreen forest.

4.4. Southern pine forest

Southeastern Evergreen Forest

In the southern pine forest, the vegetation is open with grasses and other herbs growing between the trees (fig. 4.4). Larger open spaces occasionally result from windfall, but pines are able to reseed and reclaim the openings. The trees that make up this forest are not true climax species as are those of the deciduous and boreal forests but rather are maintained by periodic burnings. Climax species are those that will not be replaced naturally by other species. If fires were eliminated, this region would eventually become oak-dominated deciduous forest. The reason it remains a pine forest is that the deciduous forest seedlings are destroyed by fire. This forest is of great economic importance as a source of timber and wood pulp for paper. Controlled burning is sometimes practiced today to assure its continuance and to keep the less-valuable deciduous species from becoming established.

This forest existed for thousands of years before Columbus made his voyages. The seasonally dead grasses growing between the trees were regularly burned in ground fires started by lightning. Ground fires have been associated with the southern pine forest long enough for the trees to have evolved features that protect them from fire damage. These include a dense tuft of needles around the growing point in seedlings and an extensive root system that results in regeneration if the growing point is damaged. In mature trees, there is a very thick outer bark that protects inner growing tissue. There are two beneficial aspects of ground fires to the southern pine forest: (1) there is a reduction in the incidences of fungal infections of trees, and (2) reseeding is favored because pine seeds germinate best on the bare mineral soil exposed after a fire. This forest will persist only as long as periodic ground fires remain a part of its ecology.

Soils of the southeastern evergreen forest are classified as red-yellow podzols, but they show the influence of another soil-forming process known as laterization. The warm humid climate in this area favors some leaching of the silicon compounds in sand. There is a more rapid rate of decomposition of organic matter and consequently less humus in the soil. The leaching of silicon compounds leaves a greater proportion of iron and aluminum in the topsoil, which may become yellow or red by oxidation. These are not fertile soils, and in order to be productive for farming they require intense conservation measures and heavy application of fertilizer.

The natural vegetation in this region consists of pine forests along the coastal plain from New Jersey to Florida and Texas. The main species of trees are longleaf pine *(Pinus palustris)*, shortleaf pine *(P. echinata)*, loblolly pine *(P. taeda)*, and slash pine *(P. elliottii)*. On inland areas, these forests grade into the oak *(Quercus spp.)* and hickory *(Carya spp.)* forests of the southern deciduous forest.

Tropical Rain Forest

Farther south in the tropics, the silicon dioxide (sand) in the topsoil is completely leached away. The remaining compounds of oxidized iron give the soil a red or yellow color that obscures the boundary between the topsoil and the subsoil. These are laterite soils that are so sterile that water percolating through them into streams is almost as pure as rainwater. Most of the mineral nutrients of this biome are in the standing forest. When leaves, twigs, and other organic debris fall to the ground, they decompose very rapidly and release their nutrients to rootlets near the surface. If the forest cover is removed and the soil exposed to the drying action of the sun, it hardens into a concrete-like substance that will never again support forest vegetation.

5

How and Where Trees Grow

How Trees Grow

Trees have three growth zones: at the tips of the stems, at the tips of the roots, and the zone between the bark and the wood. At the tips of stems and roots, the growth zones are called apical meristems. The tree grows in height by action of the apical meristems at the stem tips. The root apical meristem gives rise to extensive underground parts of a tree, which equal or exceed in length the aboveground parts. The zone between the bark and the wood is called the vascular cambium, and growth in this area increases the girth of the tree.

Growth at the apical meristem can be observed by examining the twig at the end of a branch. The terminal portion of the twig will have the leaves of the current season's growth. At the base of this section will be a series of fine rings around the twig. These are bud scale scars, and they mark the location of the bud at the beginning of the season. Farther down the twig, other series of bud scale scars can be seen marking last year's growth and the years before that. It is not unusual to see twigs showing several years of growth increments. The use of twigs to identify trees in winter is discussed in a chapter 6.

The vascular cambium consists of a single layer of cells that forms a hollow cylinder around the trunk, which is continuous with cambial cylinders around the branches and twigs. If all of the tree could be removed but the vascular cambium, the tree would be perfectly outlined by a series of hollow tubes with walls one cell thick and continuously decreasing in size from the ground up. The growth initiated by the vascular cambium can be observed by looking at the end of a log or the stump of a recently felled tree. Concentric rings of wood will be observed. Each ring represents the accumulation of wood from one year of growth and is called an annual ring. The age of the tree can be determined by counting the annual rings. Since a tree grows in diameter by the addition of wood to the inside, the outside layers burst, and the bark of an older tree is typically cracked and ridged.

In spring when the new leaves of trees appear, there is a surge of growth in which large wood cells are formed by the vascular cambium. As the season advances, often with a decrease in available water, smaller cells are formed and growth slows until it stops entirely in autumn. When large cells develop in the following spring next to the small late-summer cells from the previous season, it creates a visible line (fig. 5.1). The wood between consecutive lines constitutes the annual growth rings.

The annual rings in the central section of the trunk are darker in color and are collectively called the heartwood. The outer rings are lighter in color and consist of cells that conduct water from the roots to the branches and leaves. This section of the trunk is called the sapwood. In the heartwood, the cells have become plugged with the waste products of metabolism and no longer conduct water (fig. 5.1).

Several factors may influence the thickness of annual rings, but the most significant is water availability. In general, the greater the amount of rainfall in a growing season, the thicker the annual ring. Thus, by examining the annual rings of a tree, one can determine not only its age but also the wet and dry years of its life span. By matching groups of similar rings in overlapping years of growth in young and older trees and then extending this comparison to the timbers of older and older buildings, climatic records have been constructed spanning thousands of years. The science of using tree rings to date climatic and other events is called dendrochronology.

5.1. Heartwood, annual rings

Annual rings are characteristic of wood in both deciduous trees (angiosperms) and conifers (gymnosperms). But there is a basic difference in their woods. In deciduous trees, the wood contains two types of water-conducting cells in addition to very hard fiber cells. Deciduous trees are referred to as hardwoods. The conifers have only one type of water-conducting cell and no fibers. They are called softwoods. The beautiful patterns on untreated wood and on stained or varnished furniture are the result of the way the annual rings were cut in preparing the lumber.

The Climate for Trees

In North America there are several distinct types of vegetation, each associated with a geographic region. These are known as biomes, and four of them—western evergreen forest, boreal forest, deciduous forest, and southern evergreen forest—as their names suggest, support the growth of trees. North American forest biomes were discussed in chapter 4.

In order for trees to grow in any area, the precipitation must be equal to or greater than the potential for evaporation. For example, imagine an area where the total amount of evaporation from an open body of water, such as a pond or a lake, is 30 inches in a year. Then assume that the total amount of precipitation in that area is 35 inches a year. This region has a precipitation-evaporation ratio greater than one; it will support the growth of trees. However, this is an oversimplification because the yearly distribution of precipitation is as important as the total amount. In some parts of the midwestern grassland, there is enough precipitation in summer to support tree growth but not enough in winter. In the shrub areas of southern California, there is often enough precipitation to support tree growth in winter but not in summer. For an area to support a natural vegetation of trees, the precipitation must be 25 to 30 inches (62–75 cm) or more and be fairly evenly distributed throughout the year.

Temperature does not seem to be a factor in determining where trees can grow. They can grow in temperatures ranging from 120°F (49°C) in oases in the Sahara Desert to -70°F (-57°C) in Siberia. A factor that does exert an important influence in determining the northern limit for tree growth is duration of the growing season. During the growing season, trees must produce seeds and manufacture enough carbohydrates to last through the winter. It has been estimated that in order to do this, trees need a growing season of at least eight weeks in which the average temperature does not drop below 50°F (10°C). If the season is shorter or the average temperature lower, there are no trees. In most of eastern and northwestern North America, the only places in which trees do not grow are where they are kept out by conscious human effort or where it is too wet for them. The length of the growing season and the seasonal distribution of precipitation in these areas are very favorable for trees.

The Evergreen Forest

If nature was left to pursue its own course, undisturbed by interference from humans, all of eastern and northwestern North America would be forest. That is the way it was before Europeans arrived. In the northeast there are two types of forests: deciduous and evergreen. The deciduous forest extends from Georgia, Alabama, Mississippi, Louisiana, and eastern Texas northward into southern Canada. Beginning in central Canada and extending from the Atlantic Ocean to the Pacific is the largest evergreen coniferous forest in the world. It is known as the boreal forest or taiga, and it continues in a similar band at the same latitude across Siberia and northern Europe. The main trees of this forest in North America are balsam fir *(Abies balsamea)*, white spruce *(Picea glauca)*, black spruce *(Picea mariana)*, jack pine *(Pinus banksiana)*, and larch or tamarack *(Larix laricina)*. The combinations and proportions of these trees depend on environmental conditions in that part of the forest. For example, in wet areas black spruce and larch are commonly the most abundant trees. The boreal forest dips into the Unites States in northern New England and the northernmost parts of the Great Lakes states.

The Summergreen Forest

South of the boreal region in eastern North America is the deciduous or summergreen forest. The geographic area where conditions are best for tree growth in the deciduous forest is in the Cumberland and Allegheny Mountains. This includes parts of Pennsylvania, Ohio, West Virginia, Kentucky, Tennessee, Georgia, and Alabama. The forest in this region is called the mixed mesophytic forest. It contains thirty-five or more species of trees, the greatest number of any forest in the United States or Canada. The most abundant species are beech *(Fagus grandifolia)*, buckeye *(Aesculus flava)*, cucumber tree *(Magnolia acuminata)*, basswood *(Tilia americana)*, tulip tree

(*Liriodendron tulipifera*), sugar maple (*Acer saccharum*), white oak (*Quercus alba*), and, in protected coves and on north-facing slopes, hemlock (*Tsuga canadensis*). These occur in various combinations with other tree species in response to minor environmental variations.

In all directions from the mixed mesophytic forest, conditions for tree growth are less favorable and the number of species declines. On the eastern, western, and southern margins, it grades into forests dominated by oak: mixed oak to the east, oak-hickory to the west, and oak-pine to the south. To the north, covering much of Pennsylvania, New York, and New England and extending into Canada to the margin of the boreal zone is the hemlock-hardwoods forest. The most abundant species in this region are hemlock, sugar maple, beech, yellow birch (*Betula alleghaniensis*), basswood, and white pine (*Pinus strobus*). As the names of the forest regions suggest, they are usually named for the two or three tree species that are most abundant.

Along the coastal plain from New Jersey into Florida and along the Gulf Coast to Texas is the southeastern evergreen forest. The main species that give this forest its name are longleaf pine (*Pinus palustris*), shortleaf pine (*P. echinata*), loblolly pine (*P. taeda*), slash pine (*P. elliottii*), and Virginia scrub pine (*P. virginiana*). For tens of thousands of years before Europeans arrived, this forest was subjected to periodic fires, either started by lightning or in the last ten thousand years by Native Americans, deliberately or accidentally. As a result of settlement with more attention to fire prevention, the pines in some areas are being replaced by oak-hickory forests. This suggests that if fire was excluded permanently from the southeastern evergreen forest, the region would eventually be dominated by deciduous species.

The Ages of a Forest: Ecological Succession

Before the arrival of Europeans, eastern North America was covered with primeval forests broken only by swamps and bogs, rivers and lakes, rocky outcrops, and occasional openings caused by windfalls and fire. In the last three hundred years, most of this region has been cleared for crops or cut over for timber two or more times. Very few of those original forests remain. When traveling through the deciduous forest today, forests in all stages of maturity can be observed.

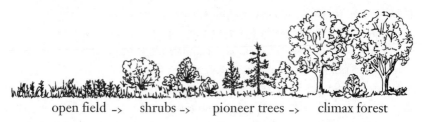

open field ->   shrubs ->   pioneer trees ->   climax forest

5.2. Ecological succession

When any tract of land is cleared for crops or other human uses and then abandoned, it immediately begins to revert to forest. The early stages of reclamation consist of a bushy field of shrubs and tree saplings. Some trees are able to rapidly disperse seeds and establish seedlings in open fields. As time passes, the trees gradually overtop and shade out the shrubs and other open-field plants. The shade-tolerant seedlings of trees such as those of the mixed mesophytic forest are then able to become established, and eventually in the mature forest, they become the most abundant trees. This whole process from open field to mature forest is called ecological succession (fig. 5.2). Succession may require hundreds of years, and the end result is the climax vegetation.

From Canopy to Forest Floor: Vertical Structure

In a mature deciduous forest, there are four distinct levels or layers, each having its own ecological significance (fig. 5.3). The highest level is the canopy layer, which covers the forest floor like a leaky green umbrella. Most of the sun's rays are blocked, but raindrops penetrate readily. The upper surface of the canopy may be 120 feet or more above the forest floor. The trunks of canopy trees are usually widely spaced, straight, and unbranched for 20 to 50 feet (6–15 m) above ground level. Shade-tolerant seedlings on the forest floor grow slowly stretching upward toward the light. When a canopy tree dies or is the victim of lightning or wind, it opens a window of light that stimulates the growth of waiting saplings, and the opening is soon closed.

The canopy layer receives the direct rays of the sun and controls the energy flow in the forest ecosystem. The environmental conditions in a mature forest are quite different from those in an adjacent open field. The for-

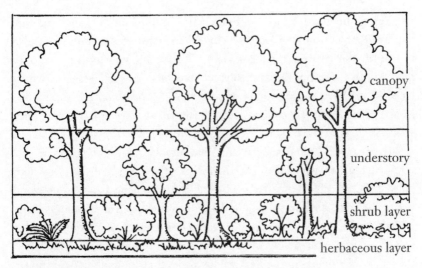

5.3. Vertical structure

est modifies and moderates the conditions of life under the canopy. Summer studies of wildflowers on the forest floor have demonstrated that the temperature near their leaves does not get as high during the day or as low at night as the temperatures near the ground outside the forest. You can judge for yourself the influence of the canopy if you walk into a forest on a hot day in mid-July. The umbrella of the canopy keeps the moisture in and reduces the movement of air. You will immediately notice a drop in temperature. You will be walking into a place that is cool, moist, and still.

The second highest level or layer of vegetation in the climax forest is made up of secondary or understory trees. These are small trees that bump against the underside of the canopy but do not become a part of it. The light that reaches this level filters through the canopy, so these trees are slow growing. They have a scattered distribution, and their trunks are usually not larger than 6 to 8 inches (15–20 cm) in diameter. Understory trees that are more common in southern forests are umbrella tree *(Magnolia tripetala)*, dogwood *(Cornus florida)*, redbud *(Cercis canadensis)*, and sourwood *(Oxydendron arboreum)*. Hornbeam *(Ostrya virginiana)*, blue beech *(Carpinus caroliniana)*, and sassafras *(Sassafras albidum)* are widely distributed, but striped maple *(Acer pensylvanicum)* grows mostly in northern forests.

Below the understory trees in a mature forest is the shrub layer. Shrubs are woody plants that usually grow in clumps of a few to many stems. They may vary in height from 1 to more than 10 feet (.3–3 m). Through millions of years of evolution, they have developed characteristics that make it possible for them to grow in the forest shade. They could not survive in an open field. Some deciduous forest shrubs are spicebush *(Lindera benzoin)*, witch-hazel *(Hamamelis virginiana)*, common elder *(Sambucus canadensis)*, maple-leaved viburnum *(Viburnum acerifolium)*, alternate-leaved dogwood *(Cornus alternifolia)*, and, in the southern forests, white laurel *(Rhododendron maximum)*, pawpaw *(Asminia triloba)*, and mountain laurel *(Kalmia lattifolia)*.

The lowest layer of plants in the forest is the herbaceous or nonwoody layer. Like the plants at the shrub level, the woodland herbaceous plants have evolved traits that contribute to their survival in an environment dominated by the canopy trees. They are often less than a foot (30 cm) high, but some may grow to a height of two or three feet (60–90 cm). Regardless of the size of the plants in the herbaceous zone, their rate of growth, time of flowering, and other aspects of their life cycles are all timed so that they coincide with events in the life cycles of the canopy trees. Examples of herbaceous plants are given in chapter 6.

6

Through the Year

An important feature of the deciduous forest that sets it apart from the boreal forest and other evergreen forests is that it has four distinct seasons. In contrast, the taiga is constantly green—summer, fall, winter, and spring. The four seasons in the deciduous forest are only loosely correlated with the calendar seasons. The latter are based on the astronomical position of the earth relative to the sun, and they divide the year into four approximately equal segments. In the forest, the seasons are marked by botanical rather than astronomical events. The canopy trees of the climax forest provide the markers for the beginning of ecological seasons. These are the seasons to which the forest and all its living creatures respond.

Autumnal Coloration

According to the traditional calendar, autumn begins about September 22 (autumnal equinox), but in some parts of the deciduous forest ecological autumn may start several weeks earlier. The daylight period begins to get shorter in late summer, and this serves as a signal to the forest trees. Certain chemical changes are initiated and the effect is that the trees begin to close down their summer activities. Substances are produced that promote the onset of dormancy and the beginning of a special layer of cells called the abscission layer between the leaves and their points of attachment to the branches. Autumn then, is a very active time, chemically, for the canopy trees and other woody plants of the forest. As you travel through the deciduous forest in autumn you will not be aware of this chemical activity, but

75

6.1. Big-leaved Aster *(Aster macrophyllus)*

you will be able to see its earliest signs, the beginning of autumnal coloration. This signals the beginning of ecological autumn, and it may begin as early as late August rather than in late September.

As autumn progresses, the special layer of cells developing at the base of the leaf first interferes with, then finally cuts off entirely the supply of water to the leaf. As this is happening, the green chlorophyll pigments disintegrate, revealing the orange and yellow pigments that were previously hidden by the chlorophyll. In the warm sunny days and cold nights of autumn, red pigments are manufactured that augment the oranges and yellows. This is the most colorful time of the year for woody plants of the forest.

This time of year is also the most colorful for some plants in the herbaceous layer. The blue to lavender flowers of heart-leaved aster *(Aster cordifolius)* and the big-leaved aster *(A. macrophyllus)* can be observed from August to October (fig. 6.1). During the same period, the bright yellow of blue-stem goldenrod *(Solidago caesia*, see fig. 6.25) and the white flowers of white snakeroot *(Eupatorium rugosum*, fig. 7.13) add splashes of color to the forest floor. White snakeroot is a plant of some historic interest (see chapter 7).

Other herbaceous plants have autumn fruits that are more colorful than their flowers. For example, white baneberry or doll's eyes *(Actaea alba)* has clusters of small white flowers in late spring. In autumn it produces white berries on bright red stalks that are much more impressive than the flowers (fig. 6.2). The berries are the source of the name doll's eyes. Blue cohosh *(Caulophyllum thalictroides)* is another plant with inconspicuous flowers in spring but large, showy blue fruit in autumn (fig. 6.3).

6.2. Doll's Eyes *(Actaea alba)*

6.3. Blue Cohosh *(Caulophyllum thalictroides)*

Plants with bright red autumn fruit include partridge berry *(Mitchella repens,* fig. 7.20), red baneberry *(Actaea rubara)*, false Solomon's seal *(Smilacina racemosa,* fig. 7.22), jack-in-the-pulpit *(Arisaema triphyllum,* fig. 6.4), wintergreen *(Gaultheria procumbens,* fig. 6.5), and ginseng *(Panax quinquefolius,* fig. 7.17).

6.4. Jack-in-the-pulpit *(Arisaema triphyllum)*

The Leaves Come Down: Winter

As the water supply to the leaf is cut off, the cells die and the bright autumnal coloration fades. In the wind and rains of October and November, the leaves break away along the abscission layer that started forming in early autumn. Soon the trees will be bare and ecological winter will have begun. The farther north one travels, the earlier ecological winter begins. In many parts of the northeast, the leaves are frequently on the ground and winter has begun by

6.5. Wintergreen *(Gaultheria procumbens)*

October 31. In these areas, December 21 (winter solstice) does not mark the beginning of anything. By this time, seeds have been dispersed and the trees are in full dormancy. For those who do not live in the northeast, ecological winter may begin later than October 31.

In early winter, the forest is characterized by a brown insulating carpet of dead leaves and the gray to brown skeletons of trees. Later in the season when temperatures are at their lowest, a layer of snow is likely in northern sections, giving additional insulation to the forest floor. The result is that in a mature deciduous forest the ground seldom freezes. This is of importance for survival because the greatest threat to forest plants in winter is not low temperatures but dehydration. Winter winds may cause excessive loss of water from buds and twigs. The unfrozen soil permits the roots of trees to continue to absorb water throughout the winter. Most of the herbaceous plants of the forest survive the winter season as underground rootstocks. They also must have a constant supply of water.

The winter forests are not entirely without color. The browns and grays are interspersed with patches of green in both the tree and herbaceous layers. The dark green, shiny leaves of wintergreen (fig. 6.5) can be seen in the herbaceous layer. These are small plants seldom more than 4 or 5 inches (10–13 cm) high, occasionally still bearing a bright red berry or two that have been missed by the birds or woodland mice. When crushed, the leaves and berries smell and taste of oil of wintergreen.

Partridge berry (fig. 7.20) is also an evergreen plant of the forest. It has a stem that trails along the ground forming mats that usually do not rise more than an inch (2.5 cm) above the ground. It can be recognized by its small, paired, roundish leaves with greenish-white veins. On the ends of branches, it bears bright red berries that may persist throughout the winter if not eaten by ruffed grouse, wild turkey, or some other forest dweller. The berries can be eaten by humans, too, but they are full of seeds and rather tasteless.

Two other plants that may be seen on the winter forest floor are hepatica (*Hepatica acutiloba* and *H. americana*) and trailing arbutus (*Epigaea repens*). Hepatica leaves have three lobes thought by someone in the past to resemble the human liver; consequently it was named hepatica or liver leaf (fig. 6.6). The winter leaves of hepatica are reddish green and are still present when the plant blooms in spring. Trailing arbutus is less common than hepatica. It is a creeping plant with oval leaves that are heart-shaped at the

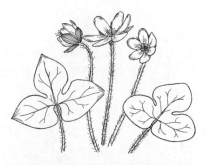

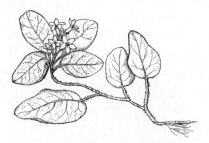

6.7. Trailing Arbutus *(Epigaea repens)*

6.6. Sharp-lobed Hepatica *(Hepatica acutiloba)*, Round-lobed Hepatica *(H. americana)*

base (fig. 6.7). It grows in rocky or sandy woods, and the winter leaves are sometimes weatherbeaten and insect damaged. Although it is protected by law, it has been extensively collected, making it rare in some regions. Trailing arbutus is the floral emblem of Nova Scotia and the state flower of Massachusetts.

Another source of color in the early winter forest is provided by witch-hazel. This plant is remarkable because it blooms during ecological winter. Witch-hazel is a large shrub that blossoms in late October, November, and sometimes into December. It has clusters of bright yellow flowers with long, stringlike, crinkly petals (fig. 6.8). The leaves and twigs steeped in alcohol are the source of witch-hazel lotion, which has been described as a tonic and a healing astringent that can be taken internally or applied externally.

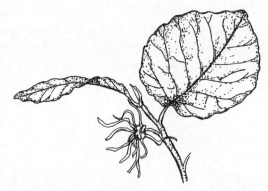

The drab colors of the winter deciduous forest are embellished by 6.8. Witch-hazel *(Hamamelis virginiana)* the presence of evergreen trees. In southern forests, pines are abundant where the deciduous forest grades into the southern evergreen forest. Farther north there is a sprinkling of evergreens that include white pine *(Pinus*

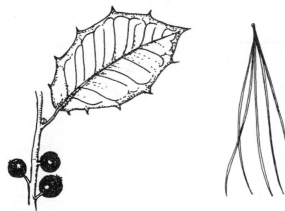

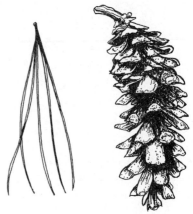

6.9. American Holly *(Ilex opaca)*

6.10. White Pine *(Pinus strobus)*

strobus),, shortleaf pine *(P. echinata)*, hemlock *(Tsuga canadensis)*, and American holly *(Ilex opaca*, fig. 6.9). White pine is the easiest of the pines to identify because it has long soft needles that grow in clusters of five (fig. 6.10). It reaches a height of more than 200 feet with a long straight trunk of up to 6 feet (1.8 m) in basal diameter. The trees were highly valued as masts in the days of sailing ships and are still valued as sources of strong, easily worked lumber. The white pine was one of the first trees to be harvested after settlement of North America by Europeans.

Shortleaf pine has needles that grow in clusters of two or three and are 3 to 5 inches (7.5–12.5 cm) long. It has a straight trunk and may grow to a height of 120 feet (36 m) or more. Shortleaf pine is a source of the valuable southern yellow pine lumber. It is one of the main species of the southeastern evergreen forest, but it ranges as far north as southern New York State.

6.11. Hemlock *(Tsuga canadensis)*

The hemlock tree attains a height of 100 feet (30 m) or more, with a straight trunk of 4 feet (1.2 m) or more in basal diameter. It is easily identified by its short flat needles that have two white lines on the undersides (fig. 6.11). In early timbering operations, hemlock was not considered an important source of lumber, but hemlock forests were cut for the bark alone. The inner bark has a high concentration of tannin, which was used in tanning leather. Hemlock bark was the main source of tannin for the tanning industry of the northeastern United States and Canada.

Identifying Trees in Winter

During summer the most commonly used feature for identifying trees are leaf characteristics. To the summer-oriented individual, the fall of leaves in autumn marks the end of deciduous tree identification. This assumption that trees in winter lack identifying characteristics is entirely false. An examination of any leafless twig will reveal some traits that are the same all year and some that are reflections of summer growth activities (fig. 6.12). The following paragraphs describe the features that are important in the identification of deciduous trees in winter. Books with winter keys are included in the references at the end of this book.

6.12. Winter twig

Buds. Buds contain, in the dormant state, the embryonic stems, leaves, and sometimes flowers of the next season's growth. In most deciduous trees, the buds are covered with one or more tiny modified leaves called bud scales that protect the embryonic parts. The number and arrangement of bud scales is usually constant for a given species. In willow *(Salix spp.)* and a few other species, there is a single cone-shaped scale, recognizable by a smooth unbroken surface. Some species such as tulip tree have two scales that fit together like the sides of a clam shell. Most commonly, as in beech, birch, maple, and oak, there are several scales that overlap like shingles. In the spring when the bud resumes growth, the embryonic stem elongates and the bud scales drop off, leaving a ring of bud scale scars. These can be used to identify the beginning of each season of growth.

For many species of deciduous trees, there is a large bud, the terminal bud, at the tip of the twig, and smaller ones along the sides. The smaller buds form in the leaf axils, the upper angles between the leaf and the stem, and are called axillary buds. Ordinarily there is a single axillary bud for each leaf in deciduous trees. In some species, the growing tip of the twig continues to grow until it is halted by cold autumn weather. The growing point then dies back to the first well-developed axillary bud, which takes over the growth in length for the next year. These species have no terminal bud.

It is very important in identification to accurately determine the presence or absence of a terminal bud. Terminal buds are usually larger than axillary buds. Since the latter form in the axils of leaves, if the uppermost bud is immediately above a leaf scar, it is axillary and not a true terminal bud. When there is no terminal bud, there is either a small stub or a branch scar, usually opposite the leaf scar, reminiscent of the true stem tip.

Leaf scars. The places of attachment of leaves on a stem are called nodes, and the spaces between nodes are internodes. As autumn advances, corklike cells of the abscission layer develop at the base of the leaf stalk or petiole. The cells on the leaf side of this layer become thin walled and the leaf eventually breaks off, leaving a scar in the shape of a cross section of the petiole base (fig. 6.12). These scars are important features in identifying winter trees. Depending on whether there are one, two, or more than two per node, they are described as alternate, opposite, or whorled. It should be recalled that there will be an axillary bud on the stem immediately above each leaf scar.

Within the leaf scars will be what appear to be small dots or raised lines. These are traces of the vascular strands that supplied the leaf with water and mineral nutrients from the vascular system of the stem. They are called bundle scars. The number and arrangement of bundle scars are sometimes very important in the identification of a winter tree. If the bundle scars are indistinct, it may help to make a thin slice across the leaf scar with a razor blade and add a small amount of an alcohol dye solution.

Stipule Scars. Stipules are small leaflike structures that are attached to the stem on each side of the leaf petiole. On winter twigs, stipule scars may be observed on each side of the leaf scars as narrow lines that extend partway or completely around the twig. Since all trees do not have stipules, the presence or absence of stipule scars may be helpful in identifying a winter twig.

Pith Characteristics. In the center of the stem is a tissue, made up of large thin-walled cells, called the pith. It can be observed by making a slanting cut across the twig with a razor blade. If the pith is solid throughout, it is called homogeneous; if it has irregular cavities with the appearance of Swiss cheese, it is spongy; if it is solid with darker cross partitions, it is diaphragmed; and if there are cavities separated by cross partitions, it is chambered. The color and shape of the pith are also diagnostic features in the identification of some species. To determine the shape, it may be helpful to apply a small amount of dye such as food coloring.

The Bursting Buds: Spring

The shortening days of late autumn stimulate the leaves of trees to produce a dormancy-inducing substance. This substance may be a chemical called abscisic acid, and before the leaves are shed it is transported to the buds. Abscisic acid and perhaps other substances are growth inhibitors. Their presence keeps the buds from bursting into growth during warm spells in winter. Before growth can be resumed in spring, the growth-inhibiting substances must be dispelled. Plant scientists are not sure how this is accomplished, but it is known that in order for the bud to resume growth, it must be exposed to an extended period of cold weather. The length of the cold period and the degree of cooling is not known for all species, but among the cultivated plants, peach trees require four hundred hours and blueberry plants eight hundred hours of exposure to temperatures less than 46°F (8°C) to break dormancy.

After a period of cold exposure, plants require increasing temperatures and increasing day length to resume growth. Most deciduous trees require a specific number of hours of warming before new growth will begin. When all of these conditions have been met, the first indication of renewed growth is a swelling of the buds. This is the signal that dormancy is broken, and it marks the beginning of ecological spring.

According to the traditional calendar, spring begins on or about March 21 (vernal equinox), but the trees in many parts of the deciduous forest have not even come close to their warming requirement by this date. In the northeast, a more reasonable date for the swelling of buds and the beginning of ecological spring is April 15 to May 1. The further south one travels, the earlier the date will be. In the southernmost part of the deciduous forest, ecological spring may almost coincide with calendar spring. During years when low temperatures persist into April and May, even though the length of the day is favorable, ecological spring may be correspondingly later. The significance of this adaptation to survival is that there is no burst of growth by trees during a warm spell in early spring followed by subsequent killing of new growth as freezing temperatures return.

The warmth generated by the direct rays of the sun plus a lengthening daylight period have the effect of an alarm clock ringing a wake-up call, not only to the trees but to the spring flowers as well. A short time later, the brown leaf-strewn forest floor will be brightened by the pale blue flowers of

6.13. Long-spurred Violet *(Viola rostrata)*

6.14. Trout lily *(Erythronium americanum)*

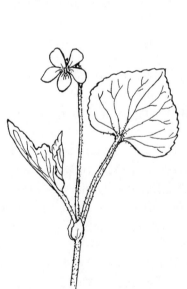

6.15. Downy Yellow Violet *(Viola pubescens)*

hepatica (fig. 6.6) and long-spurred violet *(Viola rostrata,* fig. 6.13), the yellow of trout lily *(Erythronium americanum,* fig. 6.14) and downy yellow violet *(Viola pubescens,* fig. 6.15), the deep wine-colored wild ginger *(Asarum canadense,* fig. 6.16), the white of may-apple *(Podophyllum peltatum,* fig. 7.11), bloodroot *(Sanguinaria canadensis,* fig. 6.17), rue anemone *(Anemonella thalictroides,* fig. 6.18), and white trillium *(Trillium grandiflorum,* fig. 6.19), the delicate pink of spring beauty *(Claytonia virginica, C. caroliniana,* fig. 6.20), and the deep maroon of purple trillium *(T. erectum)* and jack-

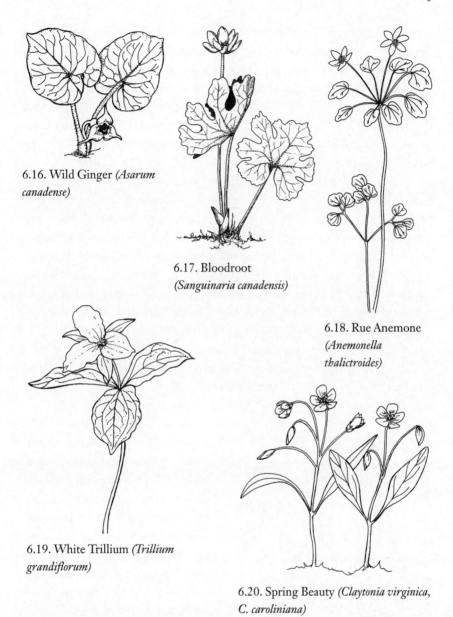

6.16. Wild Ginger *(Asarum canadense)*

6.17. Bloodroot *(Sanguinaria canadensis)*

6.18. Rue Anemone *(Anemonella thalictroides)*

6.19. White Trillium *(Trillium grandiflorum)*

6.20. Spring Beauty *(Claytonia virginica, C. caroliniana)*

in-the-pulpit *(Arisaema triphyllum*, fig. 6.4). The same alarm clock that awakened the trees and spring flowers also awakens the early insects that pollinate and thus make it possible for the spring flowers to complete their life cycles.

The direct rays of the sun strike the forest floor in spring, resulting in the highest temperature of the year. The timing mechanisms or biological clocks of the spring flowers are set for their greatest burst of growth during this period. They must accomplish two very essential activities before the leaves of the trees block the sunlight. First, if they are to survive as a species, they must produce the seeds that will become the next generation. In order to bear seeds, pollination must occur. The importance of cross-pollination was discussed in chapter 3.

The second essential activity that must be accomplished by spring flowers is to store enough food to last until the next spring season. By the process of photosynthesis during the time that sunlight reaches their leaves, carbohydrates are manufactured and stored in underground stems, bulbs, and rootstocks. After a brief period of flowering and photosynthetic activity, many of the spring flowers—including spring beauty, wild leek (*Allium tricoccum*, fig. 7.25), Dutchman's breeches (*Dicentra cucullaria*), squirrel corn (*D. canadensis*, fig. 6.21), and toothwort (*Dentaria laciniata, D. diphylla*, fig. 7.21)—will disappear from the forest until flowering time next year.

Unlike the wildflowers, most of the trees in the forest are wind-pollinated. This type of pollination is risky because air currents are variable, and there is a chance that some flowers may not be pollinated. A survival strategy of trees in response to this risk is the production of great quantities of pollen. This means that a large amount of surplus pollen is blown about by the wind, but it is a trade-off that assures the production of seeds and another generation of trees. The great quantity of wind-blown pollen is an insurance policy for the trees, but it may be a source of discomfort for some humans. Although spring is not a major allergy season, some people are sensitive to tree pollen. The pollen season starts soon after the swelling of buds and is usually over before the leaves are fully developed. Since leaves

6.21. Dutchman's Breeches (*Dicentra cucullaria*), Squirrel Corn (*D. canadensis*)

would interfere with the dispersion of pollen, the biological clocks of these trees are set for pollen release before fully developed leaves appear. Wind-pollinated trees were listed in chapter 3.

The Leaky Green Umbrella: Summer

The trees in tropical forests grow and produce new leaves throughout the year. Deciduous trees undergo a burst of growth in spring and early summer after which no new leaves appear. As the leaves develop, progressively less direct sunlight falls on the forest floor. Imagine yourself lying on your back in the forest looking upward. At the beginning of ecological spring, you will see all blue sky. As the season advances, you will see increasingly less blue until finally, when the leaves have reached maturity, you will see only their green undersides. The canopy has closed, and this marks the beginning of ecological summer. The traditional calendar indicates that summer begins on or about June 22 (summer solstice), but closure of the canopy takes place a month or more earlier. Depending on the north-south geographic location, closure may occur between early April and late May.

With closure of the canopy, there is a drop in temperature, a decrease in air movement, and an increase in humidity, creating the summer forest conditions described in an earlier section. In early summer, usually by the end of June, the buds for next spring's growth are already formed in the axils of leaves and at the tips of twigs.

Know them by leaf and bark. The number of tree species in northern deciduous forests is relatively low, making tree identification simpler for the naturalist. The long soft needles of white pine and the short flat needles of hemlock make their identification easy. Larch differs from white pine and hemlock in that it sheds its needles in winter. The needles are soft, slender, up to an inch long, and grow in dense clusters on very short or dwarf branches. In autumn they turn golden yellow prior to shedding.

Two patterns of leaf attachment can be observed among deciduous trees. The leaves may occur in pairs opposite one another, or they may occur singly showing an alternate arrangement (fig. 6.22). Deciduous trees have two types of leaves: simple and compound. Simple leaves are those in which the blade (the broad, flat part) is in one piece. In compound leaves, the blade is dissected into smaller units called leaflets. To distinguish between a leaf and a leaflet, look for the axillary bud. There is always a bud in

alternate

opposite

simple

palmately compound

pinnately compound

6.22. Leaf arrangement and type

the axil of a true leaf. The leaflets of compound leaves may be arranged pinnately or palmately. When they occur along two sides of a central stalk, it is a pinnate arrangement. When leaflets are all attached at a central point and radiate outward like fingers from the palm of the hand, it constitutes a palmate arrangement (fig. 6.22).

The trees with opposite leaves comprise a fairly small group that includes the maples, ashes, dogwoods, horse chestnut, and buckeye (the latter two are in the genus *Aesculus*). A convenient way to remember this group is to use the first letter of the first three and the first word of the fourth or fifth to form the words MAD horse or MAD buck. Anytime you encounter a forest tree with opposite leaves, you can identify it as one of the MAD horse

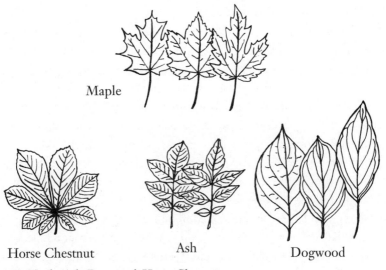

Maple

Horse Chestnut

Ash

Dogwood

6.23. Maple, Ash, Dogwood, Horse Chestnut

or MAD buck group (fig. 6.23). The maples and the dogwoods have simple leaves, but maple leaves have lobes and dogwood leaves do not (one species of maple, box elder, has compound pinnate leaves with three to five irregular leaflets and green twigs). The dogwoods are small trees or shrubs that never get large enough to be canopy trees as do the maples. The ashes have compound pinnate leaves with five to nine leaflets, while horse chestnut and buckeye have compound palmate leaves. The buckeye is a native species that grows in forests mostly south of Pennsylvania, but horse chestnut never grows wild in the forest. It is a native of Europe and Asia and is most often observed growing in lawns as an ornamental. In May it produces numerous large clusters of flowers of white petals with red and yellow bases.

The largest group of tree species in the forest consists of trees with alternate leaf arrangements. A few of these have compound leaves, including walnut *(Juglans spp.)*, hickory *(Carya spp.)*, and black locust *(Robinia pseudoacacia*, fig. 7.8). Among the trees that have simple alternate leaves are beech *(Fagus spp.)*, basswood *(Tilia spp.)*, birch *(Betula spp.)*, wild black cherry *(Prunus serotina*, fig. 7.10), elm *(Ulmus spp.)*, oak *(Quercus spp.)*, sycamore *(Platanus occidentalis)*, sassafras *(Sassafras albidum)*, and American chestnut *(Castanea dentata)*. For representative leaves of these species, see fig. 6.24.

Although trees are most often identified by their leaves, sometimes bark or twig characteristics can be helpful. Among the canopy trees, the smooth gray bark of beech cannot be confused with any other tree. This smooth surface is so inviting that many people cannot resist the destructive temptation to carve their initials. Sycamore has a mottled bark that flakes off in irregular pieces leaving a smooth greenish-gray to white surface. On shagbark hickory *(Carya ovata)*, the bark separates into long strips attached to the tree at one end and often curled outward at the other, giving the tree a shaggy look. The bark of yellow birch is yellowish in color and peels off in thin sheets around the trunk. When a twig of yellow birch is crushed, it has the odor of wintergreen.

Several species of understory trees have barks with easily identifiable features. Hop-hornbeam *(Ostrya virginiana)* has a scaly overlapping bark that resembles badly weathered roof shingles. The trunks and branches of striped maple *(Acer pensylvanicum)* have greenish-white vertical lines. The young stems and twigs of sassafras are bright green and have a fragrant odor when crushed. Blue beech or musclewood *(Carpinus caroliniana)* has a

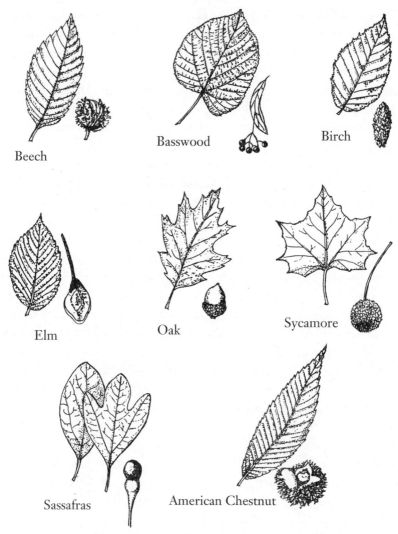

6.24. Beech, Basswood, Birch, Elm, Oak, Sycamore, Sassafras, Chestnut

smooth bluish-gray bark and a trunk with bulges that are reminiscent of strongly muscled arms.

The Forest Floor

The forest floor in summer is not as colorful as in spring. The green leaves of many of the plants that bloom in spring persist in the deep shade of the summer forest. Among these are the ones that produce showy fruits in late summer or autumn such as blue cohosh (fig. 6.3), white baneberry or doll's eyes (fig. 6.2), jack-in-the-pulpit (fig. 6.4), and false Solomon's seal (fig. 7.22). Some of the woodland plants that may be in flower from June to August are ginseng (fig. 7.17), big-leaved and heart-leaved asters (fig. 6.1), white snakeroot (fig. 7.13), blue-stem goldenrod (*Solidago caesia*, fig. 6.25), spotted wintergreen (*Chimaphila maculata*, fig. 7.15), pipsissewa (*C. umbellata*, fig. 7.16), and enchanter's nightshade (*Circaea lutetiana*, fig. 6.26).

Plants without chlorophyll. One group of summer—or autumn-blooming plants is unique in that they are not green. These plants are parasitic on either soil fungi or on the roots of trees. One of the most common is Indian pipe or corpse plant (*Monotropa uniflora*, fig. 7.19). Its waxy white color is the source of the name corpse. There is very often a cluster of stems, 4 to 6 inches (10–15 cm) in height, each with a single flower at its tip. At maturity the flower turns downward, giving the appearance of a pipe stuck into the soil stem first. As the seed capsule ages, the flower becomes erect. Pinesap (*M. hypopithys*) is a closely related species, but it differs from Indian pipe in

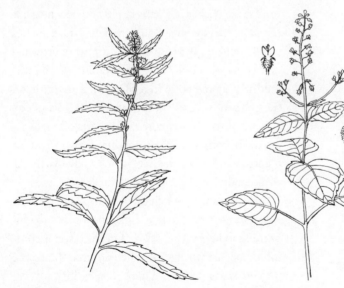

6.25. Blue-stem Goldenrod (*Solidago caesia*)

6.26. Enchanter's Nightshade (*Circaea lutetiana*)

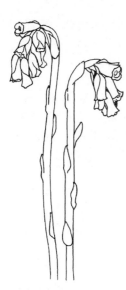

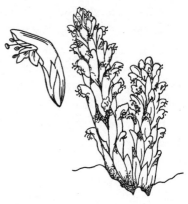

6.28. Squawroot *(Conopholis americana)*

6.27. Pinesap
(Monotropa hypopithys)

having tan or yellowish stems each with several flowers (fig. 6.27). Both of these species grow in very close association with soil fungi on which they are probably parasitic. They both turn black with age.

Two plants parasitic on the roots of trees are squawroot *(Conopholis americana)* and beech-drops *(Epifagus virginiana)*. Squawroot is 4 to 6 inches (10–15 cm) high with yellowish flowers crowded on a short stem that somewhat resembles a weathered white pine cone (fig. 6.28). It is parasitic on the roots of several species of trees, favoring oak and hemlock. Squawroot plants are often overlooked in woodlands because their brownish color blends with fallen leaves, which may partly or completely cover them.

6.29. Beech-drops
(Epifagus virginiana)

Beech-drops are light brown to purplish in color, freely branched, and may be a foot (30 cm) or more in height (fig. 6.29). The dead stems commonly persist throughout the winter and into the next growing season. There is very little difference in the appearance of dead and live stems. Beech-drops are often abundant under and parasitic on the roots of beech trees, but damage to the host plant seems to be insignificant.

Handle with care. Poison ivy (see fig. 7.1) is a woody vine most often seen in woodlands but it may grow on sand dunes, in open fields, in wet areas, or even in shady corners of your lawn. For those who are lovers of the out-of-doors, the ability to recognize this plant, with its three-parted leaf, is a must. Seventy-five to eighty percent of the population in the United States is allergic to poison ivy. The toxin is a colorless or milky fluid within special canals in all parts of the plant except the pollen. Humans are particularly susceptible to this toxin, and at least two million people develop skin rashes from exposure each year. A more detailed discussion and a drawing of poison ivy are given in chapter 7.

7

Plants of Special Interest

Early humans were much more familiar with the plants around them than the average person is today. Wild plants made up a large portion of their daily food. Medicines administered by shamans, witch doctors, or medicine men came mainly from plants. Religious rituals were often accompanied by the consumption of hallucinogenic plants. Only the medicine man needed to know the healing and ceremonial plants, but the knowledge of what could and could not be eaten was, of necessity, more widespread. Few modern Americans would survive if their lives depended on finding wild food and medicinal plants. This chapter will provide brief discussions, with drawings, of some poisonous, hallucinogenic, medicinal, and edible wild plants that may be seen on a field trip into a forest.

Poisonous Plants

Poisonous plants are those containing substances that have harmful affects on the body if they come into contact with the skin or are eaten. Considering that there are at least 300,000 species of plants, the percentage of known poisonous ones is relatively small. Nevertheless, each year hundreds of cases of poisoning are reported to the Poison Control Centers in the United States and Canada. Most of these are children who nibble on poisonous houseplants or sample the plant fare in their backyards. Poisoning in adults most often results from misidentifying a poisonous plant and using it for food or an herbal remedy.

There are no reliable physical characteristics that can be used to distinguish poisonous from nonpoisonous plants. Some writers have suggested that plants with red or white berries, milky sap, or an unpleasant odor should be avoided as potentially poisonous. To be sure, there are poisonous plants with these features, but other toxic plants have blue berries; orange, red, or colorless sap; or the pleasant odor of parsnips. To complicate matters even more, there are harmless and even edible wild plants with these same characteristics.

A belief held by many is that if one can observe other animals eating a plant it is safe for human consumption. Cattle and horses have been fatally poisoned by eating the yew plant *(Taxus spp.)*, several species of which are widely planted ornamental shrubs. These plants are also highly toxic to humans. But it is not always the case that other species respond as humans would. Although many species of birds eat the berries of poison ivy without apparent harm, it would be very dangerous for humans to consume even one. Therefore, it is not a safe practice to rely on generalities for the identification of poisonous plants. If they are being collected for human use, either as food or for home remedies, any plant that is unknown to the collector should be left where it stands.

Functions of Plant Poisons

Biologists who study plant evolution are interested in determining the origin of the physical and chemical traits of plants. Most characteristics have evolved in response to environmental conditions and contribute to the survival of the species. There is still a lot to be learned about why plants produce poisonous substances, but there are several possibilities. One is that they are waste products of metabolism. Since plants do not have excretory systems, wastes cannot be eliminated, as in animals, but must be stored in some part of the plant. Another possibility is that the poisonous substances are compounds that are essential in the normal metabolism and maintenance of the plant and their toxicity to humans and other animals is a coincidence. A third possibility is that poisonous substances have evolved as defense mechanism against their greatest natural enemies, plant-eating insects.

The heartwood of a tree (see fig. 5.1) may be a good example of meta-

7.1. Poison Ivy *(Toxicodendron radicans)*

bolic waste products being deposited in special cells where they cannot interfere with normal functioning of the plant. It is not clear whether urushiol, the poisonous substance of poison ivy *(Toxicodendron radicans,* fig. 7.1), is a waste product, an essential metabolic compound, a defensive insect repellent or none of these. It is an accident of nature that urushiol is toxic to most humans.

Types of Plant Poisons

An important and widespread group of plant poisons are called alkaloids. These are compounds that contain nitrogen and react chemically as bases rather than acids. They are almost always bitter tasting and may be present in up to forty percent of all plant families. Most alkaloids produce a strong reaction on the nervous system when ingested by animals, including humans. This action makes some of them highly toxic, but some are also very important medicinally. The names of alkaloids always end in *-ine* or *-in* and they are often named for their plant source; caffeine for *Coffea arabica,* nicotine for *Nicotiana tabacum,* and cocaine for the coca plant, *Erythroxylon coca.*

Another group of poisons that are even more widespread than alkaloids in the plant kingdom are called glycosides. Chemically, glycosides consist of at least one molecule of sugar combined with one or more nonsugar molecules. Although many glycosides are not poisonous, some are lethal. For example, cyanogenic glycosides are broken down by digestive enzymes to release deadly cyanide, which inhibits oxygen uptake by the body cells. Plants with high concentrations of cyanogenic glycosides occur in the rose and bean families. Cardiac glycosides are another group of toxic substances that act directly on the heart muscle. Lily of the valley *(Convallaria majalis,* fig. 7.2) is an open woodland plant with a high concentration of cardiac glycosides.

Other types of poisonous substances in plants are oxalic acid and ox-

alates, phenols, polypep-
tides, resins, and poisons
accumulated from min-
erals in the soil. These
compounds include the
poisons in such plants as
poison ivy, white snake-
root *(Eupatorium rugo-*
sum, fig. 7.13), and
poisonous mushrooms.
For more information on
these and other plant
poisons, see Kingsbury
(1964), Kinghorn (1979),
and Hardin and Arena
(1974).

7.2. Lily of the Valley *(Convallaria majalis)*

Types of Reactions to Plant Poisons

Allergies

It has been estimated that at least fifteen million Americans are allergic to
fungal spores and pollen and thus suffer with hay fever each year. Fungal
spores can be in the air almost any time during the growing season and even
during winter months in southern states. There are three periods during
the year when those with pollen allergies are most likely to be uncomfort-
able. The first is early spring when the wind-pollinated trees are releasing
pollen. This is usually the least severe of the allergenic periods. Trees that
may be producing pollen at this time include maples *(Acer spp.)*, elms *(Ulmus*
spp.), beech *(Fagus grandifolia)*, hickories *(Carya spp.)*, oaks *(Quercus spp.)*,
ashes *(Fraxinus spp.)*, and pines *(Pinus spp.)*. The other two allergy seasons
are associated with the flowering times of grasses and ragweeds.

Skin Irritations

The plants that most commonly cause skin irritations, or dermatitis, in
North America are poison ivy *(Toxicodendron radicans)*, poison oak *(T. pubes-*

cens), western poison ivy (*T. rydbergii*), and poison sumac (*T. vernix*). At least one of these species grows in almost every state and each of the provinces of Canada. Poison ivy is found throughout most of North America except California (fig. 7.1). Western poison ivy is widespread in the western United States and Canada. Poison oak grows in the southeast from New Jersey to Texas. Poison sumac is a shrub or small tree of swamps and marshes throughout most of eastern North America. All of these plants contain urushiol, an oily resin, to which most people are allergic. This toxin is a colorless or milky fluid within special canals in all parts of the plant except the pollen. At least two million people each year develop skin rashes from exposure to this compound.

For individuals sensitive to the toxin it might be helpful to review several facts. (1) You cannot get a reaction from simply touching a leaf or stem. The plant part must be bruised or broken so that the special canals are ruptured and the toxin comes in contact with the skin. (2) Dead leaves and stems will cause a reaction as readily as green ones. (3) The plant should never be burned because the vaporized toxin and particles in the smoke may affect the eyes, nose, and lungs.

A traditional remedy for exposure to poison ivy, poison oak, or poison sumac is to wash with strong soap as soon as possible after contact. According to the *American Medical Association Handbook of Poisonous and Injurious Plants*, this is not a good idea. It takes about ten minutes for the toxin to penetrate the skin. If a strong soap is used, the natural body oils will be removed and any remaining toxin may penetrate even faster. Since urushiol is not soluble in water, the recommended treatment is to wash with plain running water without soap.

A common misconception is that the fluid from the blisters of the dermatitis will spread the rash. When the toxin penetrates the skin, it combines chemically with deeper skin tissues. All of the toxin interacts with the cells and thus the greater the exposure, the more severe the rash. Since all of the toxin undergoes an irreversible chemical change, there is none left in the fluid of the blisters, so the rash cannot spread when the blisters burst. There are commercial lotions that claim to protect the user from the urushiol toxin. For those who are allergic, though, the best practice is to learn to recognize the plants and stay away from them.

Another source of skin irritation is from stinging hairs of plants. Wood nettle (*Laportea canadensis*, fig. 7.3) is common in moist woods throughout eastern forests and as far west as Oklahoma. Stinging hairs with tips of silica grow on the stems and the undersides of the leaves. When contact is made with the plant, the tip of the hair breaks and its jagged edges penetrate the skin. This exerts pressure on the hair, forcing fluid out of its bulblike base into the skin. The fluid contains histamine and acetylcholine, which cause an intense burning and itching sensation that may last

7.3. Wood Nettle *(Laportea canadensis)*

for an hour or more. A folk remedy for the sting of nettles is to rub the crushed stem of jewelweed or touch-me-not *(Impatiens spp.)* on the affected area of the skin.

Mushroom Poisoning

Of all the members of the plant and fungi kingdoms, the mushrooms are the most notorious as poisoners. This reputation is justified for some species, but as with green plants, only a small percentage of the total number of mushroom species are known to be toxic. There are several thousand species of mushrooms in the United States. It is estimated that about one hundred species bring about harmful reactions when ingested, and no more than ten are deadly poisonous. Although all species have not been tested for toxicity, it is believed that these estimates are not likely to change. In folklore, poisonous mushrooms are sometimes referred to as toadstools. The word is derived from a German word that means death's stool. It is not a scientific term.

As with poisonous green plants, there are no rules that a novice can follow for identifying poisonous mushrooms. The differences between edible and poisonous species are often so slight that it requires a trained expert to tell them apart. In addition, a single cluster of mushrooms may include edible and poisonous species growing side by side. This does not seem to deter persistent mushroom gatherers. Consequently, there are hundreds of cases each year of mushroom poisoning and two or three fatalities. The wild mushroom enthusiast should keep in mind a bit of pithy folk wisdom:

> There are old mushroom hunters
> There are bold mushroom hunters
> But there are no old, bold mushroom hunters.

There is great variation in the ways humans react to mushroom poisoning. No single set of symptoms can be associated with all cases. Species that cause a reaction in some people are eaten by others without problems. The degree of toxicity of a species may depend on the season it is collected, the geographic area in which it grows, or the health of the consumer. Unlike many of the poisonous green plants, the poisonous mushrooms apparently do not have a bad odor or a bitter taste. A survivor of poisoning by one of the most deadly mushrooms reported that it was delicious.

The most toxic of all the mushrooms that have been identified are species of the genus *Amanita*. Although a few species of this genus are edible, even these may be toxic to some, so the entire genus should be avoided. Most of the deaths from mushroom poisoning in North America are caused by the *Amanitas*. The names destroying angel *(Amanita verna)*, and death cap *(A. phalloides)* are appropriate descriptions of these species. A single bite of either can be fatal. The toxins from these species cause irreversible damage to the liver and kidneys. In recent years, liver and kidney transplants have saved some victims of poisoning. *Amanita* has a bulbous base from which the stalk arises, a ring of tissue around the stalk near the bottom side of the cap, and white spores (fig. 7.4). The best advice that can be offered to any amateur

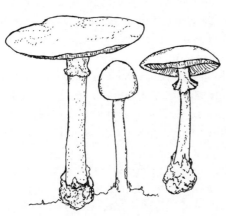

7.4. Destroying Angel *(Amanita verna)*

is to restrict mushroom consumption to those that are available from the canned goods or produce section of the supermarket.

Poisonous Plants in the Field

Fungi

Poisonous mushrooms were discussed above and hallucinogenic fungi will be discussed in a later section.

Lichens

Most lichens are harmless, but a few species in the Midwest and Northwest are known to have been responsible for the poisoning of livestock. At least one poisonous compound found in these is usnic acid. The effect of the poison is mild to severe paralysis.

Ferns

Most ferns are harmless, but there are a few species that are known to be poisonous to livestock and humans. Sensitive fern *(Onoclea sensibilis)* grows in open woods, abandoned fields, and moist meadows throughout eastern North America (fig. 7.5). The common name refers to its sensitivity to cold. It is one of the first plants to wither in autumn when night temperatures drop to the low thirties. The toxic substance is unknown, but in feeding experiments this fern has been highly toxic to horses.

7.5. Sensitive Fern *(Onoclea sensibilis)*

7.6. Bracken Fern *(Pteridium aquilinum)*

Bracken fern *(Pteridium aquilinum)* grows in open woodlands, woodland borders, and open fields, usually on dry soil. It has a brown stem, a frond with three parts, and may grow to be several feet in height (fig. 7.6). The various forms of this species are widespread throughout the northern hemisphere. It contains several poisonous substances including one that destroys thiamin (vitamin B-l) and at least two that are carcinogenic (cancer-causing substances). There have been serious cases of poisoning in cattle and horses, and the carcinogens can be transmitted to humans in milk. Although it has not been proven, eating bracken fern fiddleheads is probably a risk to human health and is not recommended.

Horsetails

Two species of horsetails, field horsetail *(Equisetum arvense,* fig. 2.8) and marsh horsetail *(E. palustre)*, have been shown to be toxic to cattle and horses. Field horsetail grows in open woods, dry and moist fields, and roadsides throughout most of the United States and southern Canada. Marsh horsetail is more frequently found in wet or moist habitats in the northern United States and southern Canada. These plants contain one of the poisons found in bracken fern, an enzyme that destroys vitamin B-1. It is not likely that humans could be poisoned by horsetails, but some herbal remedies prescribe a tea made by steeping the plant in boiling water. Such use of these plants could lead to a deficiency of vitamin B-1.

Gymnosperms or Conifers

Pine. There are numerous herbal remedies in which pine needles and resin are used to prepare tea. Most of these are harmless and some may even be therapeutic. But for the novice, distinguishing between the species of pines can sometimes be difficult. Some pines are known to have harmful effects on humans and other animals. One of these is the western yellow or ponderosa pine *(Pinus ponderosa).* This is a large tree in northwestern forests with needles in clusters of three. Eating the needles of this species is known to cause miscarriages, stillbirths, and sickly calves in cattle. In the southeastern pine forest, loblolly pine *(P. taeda)* has caused death to cattle that have eaten the needles.

Yew. There are three species of yew *(Taxus spp.)* that are widely cultivated as ornamental hedges throughout the United States and Canada, and the same number of wild species in North America. These are all evergreen shrubs or small trees. They have flat needles that usually grow in two rows along the twigs. The most common wild species in northern United States and southeastern Canada is American yew or ground hemlock *(T. canadensis).* It grows in moist to wet woods and around the margins of bogs. Reproductive structures of all the yews consist of fleshy, bright red berries, each surrounding a single seed. All parts of these plants are poisonous except the reddish tissue of the berry, not including the seed (fig. 7.7). One of the species that is used in landscaping, English yew *(T. baccata),* is considered to be one of the most poisonous trees or shrubs in Great Britain. Human deaths have occurred as a result of using yew in herbal remedies, and numerous livestock fatalities from browsing on the plant have been documented.

7.7. American Yew *(Taxus canadensis)*

Woody Flowering Plants

Black locust (Robinia pseudoacacia). This is a tree that may grow to 80 feet (25 m) in height. It has alternately arranged, pinnately compound leaves

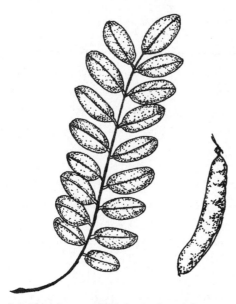

7.8. Black Locust *(Robinia pseudoacacia)*

with seven to nineteen oval leaflets (fig. 7.8). There are two thorns at the base of each leaf. It blooms from May to June with pealike, white, sweet-smelling flowers in large clusters. These give rise to long beanlike seedpods. Black locust occurs in open woods throughout eastern and central United States.

There are numerous cases on record of severe and even fatal poisoning of livestock from eating black locust leaves, bark, and twigs. Severe cases of poisoning in children, but no fatalities, have been reported for chewing on twigs, eating the inner bark, or swallowing the seeds.

This tree should not be confused with honey locust *(Gleditsia triacanthos)*, which has double pinnate leaves and branched thorns.

Buckeye and horse chestnut (Aesculus spp.). There are several species of buckeye in eastern and central North America, and one in southern Cali-

7.9. Ohio Buckeye *(Aesculus glabra)*

fornia. The eastern species *(A. glabra, A. flava, A. sylvatica, A. pavia)* are all deciduous forest trees. California buckeye *(A. californica)* is a shrub or small tree that grows at low elevations in the Pacific Coast and Sierra Ranges. The buckeye and horse chestnut have leaves that are opposite and palmately compound with five to seven leaflets (fig. 7.9).

The leaves, shoots, flowers, and seeds of buckeyes and horse chestnut contain a poisonous gly-

coside. Horses and cattle have been poisoned by eating leaves and sprouts. Children have ingested toxic, sometimes fatal,
amounts of the poison by mistaking the
seeds for those of the American chestnut
(Castanea dentata). Horse chestnut *(A. hip-
pocastanum)* is a large tree that was introduced from Europe and is usually observed
as an ornamental in lawns and gardens.

Wild black cherry (Prunus serotina). This
is a deciduous forest tree that may grow to
90 feet (28 m) or more in height throughout
eastern and central North America. In
spring it has fragrant white flowers in long

7.10. Wild Black Cherry
(Prunus serotina)

clusters that become blue-black berries in summer and fall (fig. 7.10). It has
simple alternate leaves that taper at each end with cinnamon colored hairs
on each side of the lower surface mid-vein.

This is the most poisonous of the wild cherries that grow in eastern
North America. The leaves and seeds contain large amounts of a cyanide-
producing glycoside. Large numbers of livestock of all types have died from
cyanide poisoning as a result of eating the leaves. Children have been fatally
poisoned by eating the seeds. Other cherries that contain cyanide compounds are choke cherry *(P. virginiana)*, pin cherry *(P. pensylvanica)*, and bitter cherry *(P. emarginata)*. The first two are widespread in North America
and the third grows on the Pacific Coast and east to the northern Rocky
Mountains.

Cultivated members of this genus including cherries, peaches, and
apricots, all contain cyanide-producing glycosides in their leaves and seeds.
There are cases of poisoning in children who have consumed quantities of
these seeds.

Herbaceous Flowering Plants

Doll's eyes or white baneberry (Actaea alba, fig. 6.2) is a perennial that blooms
in May and June with many small white flowers in a dense cluster on a long
stalk. In midsummer to autumn, these become a long cluster of white
berries on thick red stalks. Each berry has a prominent black dot that gives

the berry the appearance of a doll's eye. The leaves are alternate, two or three times divided on a stem reaching 30 inches (75 cm) in height. Red baneberry *(A. rubra)* is very similar but has red berries on slender stalks. White baneberry grows in much of the eastern half of North America, and red baneberry is common in the northern United States extending northward into Alaska.

All members of this genus are poisonous, and the poison is in all parts of the plant, especially the berries and roots. Ingestion of only a few berries may cause increased heart rate, irritation of the intestinal tract, and dizziness. Although there are no records of loss of life in the United States, death of children has been reported from eating the berries of European baneberry *(A. spicata)*.

May-apple (Podophyllum peltatum) is a perennial that blooms in April and May with a single large white flower in the angle between two leaf stalks (fig. 7.11). The flower gives rise to a yellow, fleshy, many-seeded fruit in July to September. The plants that do not have flowers have single, radially lobed, umbrella-like leaves on stalks that may be up to 20 inches (50 cm) high. May-apple grows in woodlands, shady meadows, roadsides, and forest margins throughout the eastern half of the United States and southern Canada as far west as Manitoba and Texas.

The ripe yellow fruit of may-apple is a well-known edible wild food. The rest of the plant and the green fruit contains a complex substance called podophyllin, which, when ingested, causes severe diarrhea and vomiting. In folk medicine, this plant has been prescribed for gall bladder problems, kidney stones, constipation, and intestinal worms. Poisoning has resulted from improper doses. As little as a tiny fraction of an ounce (5 grains or 0.324 g) of podophyllin can cause death. One of the compounds in podophyllin causes death or deformity in developing embryos. Thus it is potentially dangerous to pregnant women.

7.11. May-apple *(Podophyllum peltatum)*

Wild monkshood (Aconitumn uncinatum) is a perennial

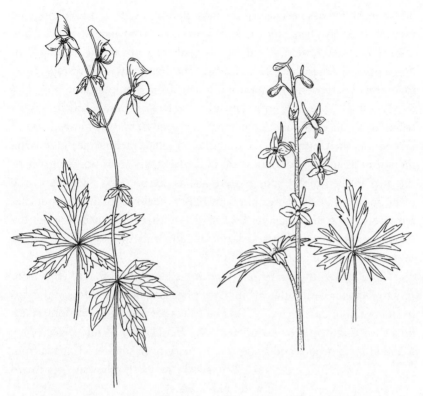

7.12. Wild Monkshood *(Aconitum uncinatum)*, Larkspur *(Delphinium tricorne)*

that blooms from August to October with five blue petals, the upper one helmet-shaped and the others narrower and unequal in size (fig. 7.12). The leaves are alternate, the lower ones long stalked with deeply cut palmate lobes. The stem is weak, often leaning on other plants, up to 4 feet (1.2 m) high but usually smaller. Wild monkshood grows in moist woods from northern Pennsylvania and southern Indiana to Georgia and Alabama. Western monkshood *(A. columbianum)* grows in the mountains from New Mexico to California and north to Canada. Garden monkshood *(A. nepellus)* is a native of Europe that is widely cultivated in flower gardens in North America.

All species of monkshood are poisonous. All parts of the plants contain several poisonous alkaloids that are so toxic as to be considered lethal. Garden monkshood has been called the most poisonous plant in England. The monkshoods have been known to be poisonous from ancient times. Most human fatalities have resulted from overdosing on herbal remedies. In North America there have been deaths, of both humans and livestock, from ingestion of parts of the plant or of tea made from monkshood.

Livestock poisoning from larkspur, a related genus, is sometimes attributed to monkshood. Species of larkspur *(Delphinium spp.)* occur throughout the United States and southern Canada in the same range as the monkshoods. The toxic compounds in the larkspurs are almost identical to the ones in the monkshoods and are just as poisonous. The leaves of the plants in the two genera are quite similar, but the flower of the larkspurs has a pronounced spur projecting backward. Another difference between the two genera is that most of the larkspurs bloom from April to July while most of the monkshoods bloom after July.

White snakeroot (Eupatorium rugosum) is a perennial that blooms from July to October with showy white flower heads in dense clusters at the tips of stems and branches (fig. 7.13). The leaves are opposite with long stalks on a stem that may grow to a height of 4 feet (1.2 m). White snakeroot is found in moist woods and along stream margins in southern Canada from New Brunswick to Saskatchewan, south to Florida and Texas.

All parts of this plant contain a poisonous alcohol known as tremetol as well as several glycosides. When the plant is eaten by almost all classes of livestock, it causes a condition called trembles that is often fatal. In cows, tremetol becomes concentrated in their milk, sometimes before symptoms of the disease are noticeable. Humans who drink the contaminated milk develop the same symptoms as other animals: tremors, nausea, jaundice, delirium, and in severe cases death. This condition, called "milk-sickness," was a common but mysterious ailment in the early 1800s, with death for up to 25 percent of those who were poisoned. In some instances, the human population of an area was reduced to one-half its original number, and whole communities were aban-

7.13. White Snakeroot
(Eupatorium rugosum)

doned. In western North Carolina there is a "Milk-sick Ridge" that is a reminder of those times. It is believed that the mother of Abraham Lincoln died of this disease. Modern processing methods have practically eliminated the danger of poisoning today.

Toxic Plant Ingestion: What to Do

Although children are most often the victims of poisoning by the ingestion of toxic plants, adults are also sometimes poisoned. The best treatment is to avoid poisoning altogether. For children, keep toxic plants out of reach. For adults, become familiar with the poisonous plants in your area and *never* consume a wild plant or mushroom unless you are *sure* of its identity. Never take an herbal remedy unless you are sure of the identity of the plants used in its construction. When a suspected poisonous plant or mushroom has been ingested, DO NOT WASTE TIME TRYING TO IDENTIFY THE SPECIMEN. GO IMMEDIATELY TO THE EMEREGENCY ROOM, OR CALL A PHYSICIAN OR THE LOCAL POISON CONTROL CENTER. If at all possible, have a sample of the suspected poisonous specimen available.

If an emergency room or a physician cannot be reached, the best practice in most cases is to induce vomiting. If a finger or blunt instrument in the back of the throat does not succeed, syrup of ipecac can be used. It contains several alkaloids that cause vomiting, and it is available as a nonprescription drug. It should be taken as soon as possible after ingestion of the suspected poisonous plant or within two hours. Ipecac should not be administered to a person who has lost the gag reflex, is not fully conscious, or shows signs of convulsions. The stomach contents should be saved, especially if a specimen of the suspected poisonous plant or mushroom is not available.

The recommended dose of ipecac for adults is two tablespoons (30 ml) and for children over one year old, one tablespoon (15 ml). These doses may vary with the individual and may not be appropriate for everyone. Ipecac should be taken with a glass of water or some liquid other than milk. For children under one year of age, ipecac should be administered only under the direction of a physician. Most people are within at least telephone distance of qualified medical assistance. The instructions presented here are under no circumstances to be followed rather than calling for medical assistance.

Hallucinogenic Plants

Definitions and Beginnings

Hallucinogenic plants contain compounds that act on the central nervous system. They bring about changes in mood and distort the ways in which time, space, color, and sound are perceived. These departures from reality are called hallucinations. Most of the known hallucinogenic plants are in the dicotyledon group of the angiosperms or flowering plants. There are none in the gymnosperms, ferns and fern allies, mosses and liverworts, or algae. In the fungi, however, there are several species that contain hallucinogenic substances.

Humans have known about and used hallucinogenic plants for thousands of years. It is interesting to speculate on how early man may have learned to distinguish between poisonous, medicinal, and hallucinogenic plants. In the learning process, no doubt, many became ill and probably many died. Then, as now, some individuals were more perceptive than others, and these became the shamans or medicine men. In their earliest uses, the hallucinogenic substances served as a means of communicating with the spirit world for guidance in times of crisis. Every known hallucinogenic plant has a history of such use in early cultures.

Consulting with the spirit world was a very serious matter in ancient cultures as well as in some modern primitive cultures. It was undertaken solemnly and often with elaborate ceremony. In his visions, the medicine man would look for answers to religious, medicinal, social, or military problems. Some writers have suggested that the very concept of God originated in these visions. During these times, the use of hallucinogenic substance was limited mainly to the medicine men and never a practice among the common people. Only in relatively recent times have these substances been subjected to widespread recreational use and often abuse.

Basing the decision solely on the compounds they contain, it is sometimes difficult to distinguish between poisonous, medicinal, and hallucinogenic plants. In this chapter, plants are arbitrarily listed in one or the other of these categories only for purposes of discussion. The poisonous compounds in some plants are important medicines when taken in controlled doses. Some hallucinogenic substances in plants are important medicines when taken at one level, but are deadly poisonous when taken in larger doses. Thus, hallucinogenic plants are not described here to suggest experimentation. One needs only to read the daily newspaper for accounts of fatal overdoses with hallucinogenic substances.

Fly agaric mushroom (Amanita muscaria). The cap is 3 to 8 inches (7.5–20 cm) in diameter with a stalk 4 to 8 inches (10–20 cm) high (fig. 7.14). The stalk has an enlarged bulbous base with several concentric rings of tissue above the swollen portion. There

7.14. Fly Agaric Mushroom (*Amanita muscaria*)

is usually a ring of tissue around the upper part of the stalk beneath the cap. The cap is dome-shaped when young, then becomes flat with age. It is usually covered with whitish or yellowish flecks of tissue. Along the Pacific coast the cap is scarlet red, and in eastern forms it is yellow-orange. The spores are white. Fly agaric mushroom grows in coniferous and deciduous forests throughout North America.

The common name, fly agaric, refers to the long-standing observation that when flies feed on this mushroom, they die. A more recent observation suggests that it simply renders them unconscious for a period of time. This is one of the oldest known hallucinogens and it may be the best known mushroom in the world. In children's literature and other writing, when a symbolic illustration of a mushroom is needed, it is usually a fly agaric that is shown.

The hallucinogenic properties of the eastern yellow-orange form are questionable, but the scarlet red form, also grown in Asia, has been used as a hallucinogen for centuries. Ancient writings in India that are more than three thousand years old refer to a sacred plant that is believed to have been fly agaric. A remarkable quality of this mushroom is that the hallucinogenic substance passes through the body and emerges unchanged in the urine.

There are reports that in early Asian religious ceremonies, participants drank the urine of one who had eaten the mushroom in order to share his visions.

Medicinal Plants

Their Importance

The shamans and medicine men of primitive cultures were probably the first professional men. Since most of the medicines they dispensed came from plants, they of necessity were botanists. This strong bond with plants by those that practice the healing arts has been a characteristic of human societies from prehistoric times. It continued into modern times until near the end of the 1800s. Even at that date, many medical doctors were botanists and most professional botanists were physicians. In 1900 about 80 percent of the drugs prescribed by physicians came directly from plants. The growth of organic chemistry, beginning at about this time, initiated an era of synthetic medicines. Although the development of synthetic medicines has continued into present times, 35 to 40 percent of all prescribed drugs are still either natural plant compounds or plant compounds in combination with synthetic substances.

Thus, plants are still as important in the practice of medicine as they were to the shamans and medicine men. Today they serve the medical profession in at least three ways. First, almost 25 percent of the drugs prescribed by modern physicians come directly from plants. The most effective known drug for the treatment of malaria, the all-time number one killer of humans, is quinine. It is derived from the bark of a small tropical tree, the quina quina tree *(Cinchona officinalis)*. Ephedrine is a drug used to treat asthma, hay fever, and colds. It is derived from an Asiatic shrub *(Ephedra sinica)*, but there are species of this genus in southwestern United States.

Another way that plants are useful in modern medicine is that some plant compounds are used as essential components in the manufacture of medicinal drugs. For example, two tropical climbing vines of the yam family in the genus *Diosocorea (D. floribunda, D. composita)* contain compounds called steroids. These have become important as essential components in the manufacture of cortisone, human sex hormones, and birth control pills.

Cortisone is used in the treatment of Addison's disease, allergies, and arthritis. The yam family mentioned above is not related to sweet potatoes. In southeastern United States, sweet potatoes are erroneously called yams. The sweet potato is actually a member of the morning glory family.

A third use of plants in modern medicine is that natural plant drugs may serve as models for the synthesis of identical or similar drugs. With the highly publicized recreational use of cocaine, its use as a medicine is sometimes overlooked. It is an alkaloid derived from the leaves of the tropical shrub coca (*Erythroxylon coca*). In the past, it was used extensively as a local anesthetic for surgery of the eyes and for dentistry. Today it has been almost completely replaced by a very similar man-made drug, procaine, better known by the brand name Novocain.

In herbal medicine, the use of willow bark to alleviate pain and reduce fever has been known for two thousand years or more. This effect is mainly the result of a glycoside, salacin, which was isolated and identified in the nineteenth century. Salacin can be converted into salicylic acid. In 1899 salicylic acid was combined with acetic acid to form acetylsalicylic acid. This substance was given the name aspirin and is probably the most widely used medicine in the world.

Only a very small percentage of the thousands of known species of plants have been chemically analyzed for medicinal drugs. One can only speculate as to the value of the medicines that remain to be discovered. It is clearly of great importance to maintain habitats for the survival of wild plants in such areas as national, state, and municipal parks, wildlife preserves, and wilderness areas. Unfortunately, the places where more than 60 percent of all plant species grow, the tropical rain forests, are the areas that are being destroyed at the greatest rates. Between 1980 and 1990, about 8 percent of all the world's tropical forests were lost. It has been estimated that in these areas about fifty thousand species each year, both plant and animal, are either being exterminated or reduced to levels at which extinction is almost certain. Each species lost to extinction represents the loss of a potential life-saving drug.

Herbal Medicine

In the early days of colonization in North America, physicians and hospitals were few or nonexistent. The settlers had no choice but to rely on herbal medicine for treating illness and injuries. They had brought with them a rich heritage of herbal remedies from Europe, and they soon added to these by including treatments learned from Native Americans. The result is that there are folk remedies for almost every ailment experienced by humans.

A complete description of all the plants, and the uses that have been made of each in herbal medicine, would require a large book. Very few of these remedies have been subjected to controlled testing to verify their effectiveness. Some can be traced to the ill-conceived doctrine of signatures, which held that the shape of a leaf, root, or seed determined its use in healing. Other herbal remedies can be traced to a time when magic and mysticism were associated with certain plants. Some folk remedies do make use of plants that contain powerful medicinal drugs—some so powerful, in fact, that a misjudged dose could result in death. Consequently, most modern physicians view herbal medicine as little more than quackery. There is an abundance of justification for this attitude, but to disregard all herbal medicine runs the risk of throwing out the baby with the bathwater. Modern medicine has its roots in folk medicine, and there may be information that it can still impart.

Poisonous plants are often components of herbal remedies. Sometimes the only feature in the use of a plant that distinguishes it as medicinal or poisonous is the size of the dose. Experienced practitioners of herbal medicine can usually recognize the signs of acute poisoning, but subtle symptoms from repeated exposure to small doses of a toxic plant drug may not be so easily recognized, even by an experienced herbalist. Modern laboratory techniques are usually necessary to detect damage to internal organs such as the liver or kidneys. However, in many parts of the world, especially in underdeveloped countries, herbal medicine is the chief source of treatment for all human ailments. In a study by the World Health Organization, it was concluded that the only way developing countries can achieve minimum health needs is to make use of traditional folk medicine.

In China, where herbal medicine has been practiced for thousands of years, there has been a fusion of folk treatments with modern methodology. Instead of treating a patient specifically for a sore knee, a liver problem, or a skin disorder, Chinese medicine prescribes for the total health of the body. A more open exchange of information between Chinese and western medicine would probably result in improvements in both.

Medicinal Plants in the Field

It is possible there are effective folk remedies needing only to be clinically tested.
It is probable too that there are some that
should be discontinued. Modern research
is constantly providing new information
with which to evaluate herbal remedies.
For example, sassafras tea, made from the
bark of the roots of the sassafras tree *(Sassafras albidum)*, has been used for hundreds of years as refreshing root-beer
flavored tea and as an herbal treatment
for various ailments. Recent studies have
revealed that sassafras oil causes cancer
in rats and is a potential carcinogen for
humans.

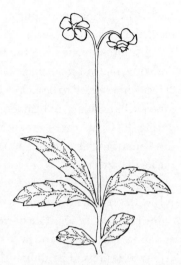

7.15. Spotted Wintergreen
(Chimaphila maculata)

The following is a representative group of plants that have been used in
herbal medicine. These are not offered as recommendations for home
remedies. Unless the user is thoroughly familiar with the plant components, most herbal remedies are inadvisable. For
the inexperienced herbalist, the local drug store
can provide better and safer medicine.

Spotted wintergreen (Chimaphila maculata). An
evergreen perennial that blooms from June to
August with waxy, fragrant, pink to white flowers
in clusters of one to five at the tip of each stem
(fig. 7.15). The leaves are in whorls of three to six,
with scattered teeth, and whitened along the
veins on the upper surface. The stem is unbranched and about 10 inches (25 cm) high. A related species, pipsissewa *(C. umbellata)*, is similar
but without white markings on the upper surfaces
of the leaves (fig. 7.16). Spotted wintergreen
grows mostly in eastern North America, but pipsissewa is found from Nova Scotia to Alaska,

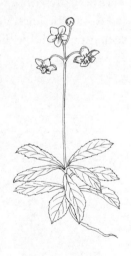

7.16. Pipsissewa
(Chimaphila umbellata)

south to Georgia and California. Both plants have similar uses in herbal medicine.

A solution made from the fresh plant is recommended to increase the flow of urine and for beneficial influence on the urinary tract. The name "pipsissewa" is said to be a Cree Indian term referring to the use of the plant for dispelling kidney or bladder stones. A tea made from the leaves was used during the Civil War as a tonic and a treatment for rheumatism. Some Native American tribes crushed the whole plant and applied it to swellings of the legs and feet.

Ginseng (Panax quinquefolius). A perennial that blooms from June to July with very small greenish-white flowers in a flat-topped cluster at the tip of the plant (fig. 7.17). The flowers become a cluster of bright red berries in autumn. The leaves are in a whorl of three, each with five palmately arranged leaflets. The unbranched stem may be 16 inches (40 cm) high. Ginseng grows throughout eastern North America and as far west as Manitoba and Oklahoma.

This plant has been used by the Chinese for thousands of years for a variety of ills and as an aphrodisiac. Although the American species is an acceptable substitute with similar medicinal properties, it has never achieved the level of use in the western medicine as it has in Asia. In herbal medicine it has been prescribed for convulsions, nervous disorders, colds, headaches, and to stop the bleeding of wounds. Japanese and Russian scientists have demonstrated that without question it is a tonic and a stimulant. It has been collected and sold for exportation almost to the point of extinction in some areas. Let it stand! Its medical functions can be better served by the local drug store.

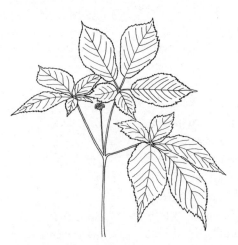

7.17. Ginseng *(Panax quinquefolius)*

Golden seal (Hydrastis canadensis). A perennial that blooms in April and May with greenish-white flowers consisting entirely of numerous long stamens and pistils (fig. 7.18). There is usually a single long-stalked lower leaf and two alternately arranged upper leaves. Each leaf has five to seven deep palmate lobes. The stem is hairy and may be 15 inches (38 cm) high. Golden seal grows in the eastern United States and southeastern Canada as far west as Manitoba, Minnesota, and Arkansas.

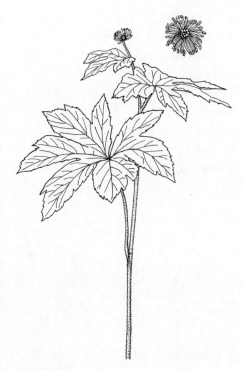

The common name refers to the rootstock, which was used by Native Americans as a source of yellow dye. In herbal medicine it has been used for inflammation of mucous membranes, hemorrhoids, and for disorders of the stomach and liver. The dried and powered root was dusted on open cuts and sores to enhance heal-

7.18. Golden Seal *(Hydrastis canadensis)*

ing. Some tribes mixed the powered root with bear grease to make an insect repellent. Like ginseng, this plant has been over-collected and will probably need protection to survive.

Indian pipe (Monotropa uniflora). A fleshy, white, sometimes pink-tinged plant that blooms from June to September (fig. 7.19). This non-green plant has a mat of fibrous roots closely associated with, and parasitic on, a soil fungus. The flowers are waxy-white or pale pink, nodding downward at first, then becoming upright as the seed capsule matures. The leaves are alternate, scalelike, and white. The stems are often numerous and may grow to 10 inches

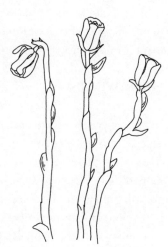

7.19. Indian Pipe *(Monotropa uniflora)*

(25 cm) in height. The whole plant turns black with age. Indian pipe grows throughout the United States and Canada.

Other common names for the plant are convulsion-root and fits-root. As these names suggest, this plant has been used to treat spasms and other nerve disorders. In early America the juice of the stem was used as an eye lotion.

Partridge berry (Mitchella repens). An evergreen perennial that blooms from May to July with white or pink four-lobed flowers in pairs at the ends of branches (fig. 7.20). Each pair of flowers becomes a red berry after the

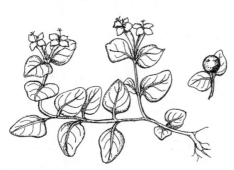

7.20. Partridge Berry *(Mitchella repens)*

flowers fade. The leaves are opposite, small, roundish, often with greenish-white veins. The stem creeps along the ground, sometimes forming mats, and may be a foot (30 cm) long. Partridge berry is found from Newfoundland to Manitoba, south to Florida and Texas.

This plant has been used as a tonic, an astringent, and to increase the flow of urine. Another common name for it is squaw vine, referring to a time when the women of several tribes of Native Americans used a tea made from the plant in the last month of pregnancy to make childbirth easier. Some tribes used the tea as a treatment for insomnia.

Edible Wild Plants

In hunter-gatherer societies, humans were essentially vegetarians. This was especially true before the invention of the spear and the bow and arrow. Even after these tools were invented, plants still made up most of the human diet. The species that were used for food by the hunter-gatherers are still growing in those areas, but very few, if any, are important sources of human food today. Instead the cereal grains—wheat, rice, corn, oats, barley, and millet—are the main food plants in the modern world. The edible native plants in any region of the earth are better adapted for survival in those climatic conditions and are sometimes more nutritious than imported cereal grains. Thus, the cultivation of wild food plants locally may offer a partial solution to escalating food problems. Exploration of this possibility is an appropriate direction for agriculture of the future.

Why Know Them?

In this high-tech era of rapid transportation and efficient freezers, when well-stocked supermarkets are available to almost everyone, who needs to know about edible wild plants? For the purposes of survival, probably no one. Even if all transportation and freezing facilities failed and supermarkets had empty shelves, knowledge of edible wild plants would be of little value to residents of New York City, Philadelphia, Chicago, or Los Angeles. There simply are not enough edible wild plants out there to feed so many people. In the event of total failure of the supply system and electricity, millions would starve to death. When hunter-gatherers foraged for edible plants, the population was measured in tens of individuals per hundreds of square miles rather than in the millions.

The most valid reason for learning to identify edible wild plants is probably the same reason that people climb mountains: just because they are there. Those who love the out-of-doors find satisfaction in being able to recognize poisonous, medicinal, and edible plants. On the practical side, it is always possible that a camper or hiker could become lost in the wild. Knowing some edible plants could be very helpful. In addition, being able to occasionally prepare a complete meal with wild plants is a novelty that is sure to surprise and perhaps delight dinner guests. Euell Gibbons (1966) has described elaborate wild food dinners in his home at which every dish was a conversation piece.

Nutrition and Taste

In colonial times, wild plants were commonly used in the home for herbal remedies and as part of the diet. As food plants, these often included the new shoots or leaves of clintonia *(Clintonia borealis)*, waterleaf *(Hydrophyllum spp.)*, wild leek *(Allium tricoccum)*, and wood nettle *(Laportea canadensis)*, collected in the spring and cooked as greens. This type of foraging for wild plants is still practiced in some rural areas in North America.

Not only are they often more nutritious, but wild plants can also be pleasing to the taste. The person who judges wild plant foods by whether or not they taste like cultivated plants will be disappointed. Just as potatoes and corn are from different plants and have different tastes, so each wild species has its own unique flavor.

As a word of caution, the wild food plant enthusiast should be wary of collecting along roadsides. Before the advent of unleaded gasoline, roadside plants were found to have elevated concentrations of lead. In areas where leaded fuel is still used, plants will continue to be contaminated. Plants growing beyond about sixty feet from the highway should have normal concentrations of lead.

What Part Should Be Eaten?

When a plant is characterized as edible, it does not necessarily mean that the whole plant can be eaten. No one seriously questions the potato as an edible plant, but there have been fatalities in humans and livestock from eating the leaves. Even the tuber may contain a toxin if it has been exposed to the sun: there will be a green surface layer that contains the same toxic substance that is found in the leaves. In the rhubarb plant, the leaf stalks are the edible parts. The blade of the leaf (the flat green part) is highly toxic, and human deaths have resulted from its consumption. The fruits of apple and peach trees are delicious and nutritious foods, but fatal cyanide poisoning can result from eating quantities of seeds of either.

In wild plants, as in cultivated ones, it is important to know what part is edible. For example, the ripe yellow fruit of may-apple *(Podophyllum peltatum)* is safe to eat in moderate amounts, but the entire plant, including the fruit when it is green, contains a highly toxic substance. Some plants have more than one edible part. A beginning student of botany should never collect wild food plants unless accompanied by an experienced collector. Sometimes poisonous plants are similar to nonpoisonous edible ones. No plant should be eaten unless the collector is sure of its identity.

Plant Conservation

It is especially important to consider plant conservation when discussing the collection of wild food plants. The individual plants of a given species are seldom randomly distributed throughout their geographic range. Instead, they often occur in scattered clumps in those parts of the range where environmental conditions are suitable for their growth and reproduction. In collecting enough plants for a single meal, an entire local colony could be eliminated. This is less damaging when the plants are perennials and are cut at ground level, leaving the rootstock to generate new plants. If the plants are annuals, collection is more damaging because young shoots, before they produce flowers and seeds, are usually the most desirable for food.

The greatest proportion of edible wild herbaceous plants are found in open spaces such as abandoned fields and farms. They are usually plants that have high rates of growth and reproduction so that the risk of extinction is less than for those of woodlands. Forests are more limited in size and number than are open spaces, and in some regions they are actually shrinking. Collecting food plants in these areas should be undertaken with care. The collector should never take more than a few plants from each colony.

Edible Wild Plants in the Field

The following descriptions and drawings of woodland plants are presented for identification only. Recipes for the preparation of wild food plants can be found in Gibbons (1962, 1966, 1979), Krochmal and Krochmal (1974), and Peterson (1979). These books are listed in the bibliography at the end of this book.

Cut-leaved toothwort (Dentaria laciniata). A perennial that blooms from April to June with white flowers in clusters at the top of the stem (fig. 7.21). The leaves are in a single whorl of three, each leaf with three slender, toothed segments. The stem is smooth, unbranched, up to 14 inches (35 cm) high. A related species, two-leaved toothwort *(D. diphylla)*, is very similar but has a single pair of leaves, each with three rounded leaflets having coarse teeth (fig.

7.21. Cut-leaved Toothwort *(Dentaria laciniata)*, Two-leaved Toothwort *(D. diphylla)*

7.22. False Solomon's Seal *(Smilacina racemosa)*

7.21). The toothworts grow from New Brunswick to Manitoba, south to Georgia and Alabama.

The edible part of the toothworts are the rootstocks. They can be eaten raw, added to salads, or ground in vinegar as a substitute for horseradish. The toothworts are woodland wildflowers and should be collected as food only in emergencies.

False Solomon's seal (Smilacina racemosa). A perennial that blooms from May to July with numerous small, white flowers in a branched cluster at the tip of the stem (fig. 7.22). These become a cluster of first green and then red berries. The leaves are alternate, pointed, and short stalked, with prominent parallel veins. The stem zigzags between leaves and grows with a slight arch rather than standing erect. It may be 3 feet (90 cm) long. False Solomon's seal ranges from Nova Scotia to British Columbia, south to Georgia, Mississippi, and Arizona.

The young shoots can be used in salads or cooked like asparagus. The rootstocks were used for food by some Native American tribes after soaking them in lye and parboiling. The berries are edible but should be eaten sparingly because they may cause diarrhea. This is a woodland wildflower and should be collected for food only in emergencies.

Hairy sweet cicely (Osmorhiza claytonii). A perennial that blooms from May to June with tiny white flowers in clusters at the ends of stems and branches (fig. 7.23). The leaves are alternate and twice divided with toothed leaflets. Stems are soft-hairy, to 3 feet (90 cm) high. A related species, smooth sweet cicely *(O. longistylis)*, is very similar but with hairless stems. The sweet cicelies grow from southern Canada south to Georgia, Texas, and Colorado.

The roots of the sweet cicelies contain anise oil and are edible. In some species of this genus, the root is too strongly flavored to be eaten, but these can be used for seasoning.

Virginia waterleaf (Hydrophyllum virginianum). A perennial that blooms from May to July with white to lavender flowers in a dense cluster at the ends of the stems (fig. 7.24). Long stamens extending from the flowers give the cluster a fuzzy appearance. The leaves are

7.23. Hairy Sweet Cicely *(Osmorhiza claytonii)*

alternate with five to seven deeply divided pinnately arranged lobes. The upper part of the stem is hairy and it may be 28 inches (78 cm) high. A closely related species, broad-leaved waterleaf *(H. canadense)*, is similar, but the stem has few hairs and leaves have five to nine palmately arranged lobes (fig. 7.24). The waterleafs grow from Quebec to Manitoba south to Georgia and Alabama.

The young leaves and stem tips, collected before flowering, can be cooked as greens in one or two changes of water to dispel bitterness. In colonial New York, these plants were cooked and referred to as John's cabbage.

Wild leek or ramp (Allium tricoccum). A perennial that blooms from June to July with white flowers in a

7.24. Virginia Waterleaf *(Hydrophyllum virginianum)*, Broad-leaved Waterleaf *(H. canadense)*

7.25. Wild Leek *(Allium tricoccum)*

globular cluster at the tip of a leafless stem (fig. 7.25). There are two or three leaves that arise from an underground bulb. The flowering stem may be 6 to 20 inches (15–20 cm) high. The leaves appear early in the spring, then wither before the flowering stem appears. The whole plant has a very strong onion odor. Wild leek grows from New Brunswick and Quebec to Manitoba south to North Carolina and Tennessee.

Wild leek is the most edible of the wild onions. The leaves and bulbs can be cooked as a vegetable or used in salads and for seasoning. Ramp banquets are a spring tradition in southern Appalachia but overconsumption can cause a halitosis problem that a breath deodorizer will hardly influence.

8

Naming and Collecting Plants

Plant Names

There are more than 300,000 known species of plants on earth today. Each of these has a name and some have more than one. When a new species is discovered, the individual who recognized it as new has the honor of giving it a botanical name. Every species thus has a Latinized botanical name. It is in Latin because this language is no longer used by human culture and it will never change. As a result, the Latin name will have the same meaning five hundred years from now as it does today. In addition to the botanical name, many plants have one or more common names that usually date to antiquity. These were given by people who were familiar with the plant because it was harmful to humans in some way, was a source of medicine, was useful for food, or had outstanding physical characteristics.

In order for a plant name to be of scientific value, it must be the one and only name that universally refers to that plant. The woodland plant wild geranium has also been called cranesbill, alum root, and chocolate root. If a botanist wishes to publish a paper on research he has conducted on wild geranium, he must use a name that will be recognized by other botanists. If he uses one of the American common names it would mean nothing to botanists in Russia, India, or Germany. This is why research reports always identify plants by their botanical names. The botanical name for wild geranium is *Geranium maculatum*, and it is spelled the same way in every language of every country in the world. No other plant on the globe has this name.

Common or Folk Names

While every known plant species has a Latinized botanical name, most do not have common or folk names. Possible reasons for this are that many plants are small and escape notice, or are growing in areas infrequently visited by humans. For example, mosses and liverworts are inconspicuous low-growing plants often collectively referred to as bryophytes (see chapter 2). There are more than 22,000 different species of these plants, some of which are difficult to distinguish even by professional botanists. Among the flowering plants, some closely related species are so similar that an entire cluster of species, called a genus, may be known by a single common name. For example, in the mint family, mountain mint is the only common name for at least fourteen species.

Although common names are unsuitable for the identification of plants in research papers, they represent a wealth of information and folklore. Most common names are descriptions of some features of the plant. For example, "blue bell" is a name that refers to the color and shape of the flower. Even though there are at least two plants that have this name, no one will be surprised to find they both have flowers shaped like little blue bells. Common names frequently refer to color; it is easy to distinguish between red baneberry and white baneberry. Blue cohosh is a woodland plant with prominent blue berries in late summer and fall. Beadruby is a name sometimes used for false lily of the valley that refers to its translucent red berries.

Other descriptive characteristics featured in folk names are odor, habitat, and geography. Stinking Benjamin, one of the names for purple trillium, is so called because the flowers have the odor of rotting meat. Wild stonecrop is a small herbaceous wildflower that grows in rocky woodland habitats. Oswego tea and Virginia creeper were so named not because these are the only places they occur, but rather because these are the places they were first observed or collected.

One of the most common themes for folk names is the medicinal uses of plants. Such names as rheumatism root, nerve-root, feverwort, and abscess root leave little doubt about the medicinal condition to which they refer. A sixteenth-century physician named Paracelsus popularized a concept known as the doctrine of signatures that had a great influence on the naming of plants. According to this concept, the creator gave each plant species a characteristic to indicate its use for humans. Many common names in use today date to this period. Liver leaf or hepatica has a three-lobed leaf that is supposed to resemble the liver; it was used as a remedy for liver problems. Bloodroot exudes an orange-red juice from its rootstock and was recommended for ailments of the blood. None of the treatments suggested by the doctrine of signatures have been substantiated by modern medicine.

Other common names are intriguing, colorful, and sometimes ominous. Enchanter's nightshade, a common woodland plant, was named for a beautiful enchantress in Greek mythology who had the power to turn men into beasts.

Botanical Names

Botanists have been using Latin to name plants for hundreds of years. Before the mid-eighteenth century, these names often consisted of several words and were more like descriptions than names. In 1753 a Swedish botanist named Carl Linnaeus wrote a book entitled *Species Plantarum*, which, freely translated, means "The Species of Plants." In the book he used a two-word system to give names to all the plants in the world known to him. Although it met with resistance from some botanists of the time, this method greatly simplified the naming of plants. In older systems of naming, catnip had been called *Nepeta floribus interrupte spicatus pedunculatus*. In the method used by Linneaus, it became simply *Nepeta cataria*. This two-word or binomial system is used by botanists today.

There is a well-defined procedure for giving a newly discovered plant species a name. Following the system initiated by Linnaeus, the botanical name consists of a generic name and a specific name. The botanists coming after Linnaeus added a third component to the name: the initials of the botanist who named the plant. All the plants named by Linnaeus have an L. following the specific epithet. For example, the botanical name for wild geranium is *Geranium maculatum* L. This plant is in the genus *Geranium*, its specific name is *maculatum*, and it was named by Linnaeus. In nontechnical publications, the initials of the botanist who named the plant are often omitted. In writing the botanical name, the genus is always capitalized and the specific epithet is lowercased. (The word "species" is both singular and plural as illustrated in the following sentence: "The genus *Geranium* has at least two species, but some genera have only one species.")

The botanical name provides both descriptive information about the species and information on its evolutionary relationships. All the species that make up a genus evolved from a common ancestor; thus a genus is a group of closely related species. Likewise, a group of closely related genera make up a family, and a group of similar families is an order. All of the genera in the lily family evolved from a common ancestral genus and are more similar to one another than they are to the genera of any other family. Most generic names are hundreds of years old and are derived from ancient Latin or Latinized Greek words. Some generic names widely used and recognized as common names by the general public are aster, geranium, hibiscus, iris, phlox, and rhododendron.

The second word of the botanical name is the specific epithet. It is usually a word that describes some characteristic of the species. For example, *Quercus alba* is the botanical name for white oak. *Quercus* is the name used in ancient Rome for oak and *alba* is Latin for white. It refers to the whitish bark and leaf undersides of white oak. It should be noted that the botanical name for this plant must include both the genus and the specific epithet. *Alba* is the specific epithet for several plants. Only when it is used with *Quercus* does it mean white oak. This is true for all specific epithets: they are parts of the botanical names only when used with genus names.

Although botanical names are indispensable for professional botanists, for the uninitiated they sometimes seem long and difficult to pronounce. *Polystichum acrostichoides* is the botanical name for the common woodland Christmas fern. Many—perhaps most—botanical names are shorter than this, and with experience and a little effort they become much easier to use. One of the pleasures of being familiar with plants is having the ability to talk with others who have similar interests. For maximum communication with others, at all levels, learning the botanical as well as the common name is recommended.

What Is a Species?

"Species" is a term that has been used frequently in the preceding pages, so a few words of explanation are appropriate. The species is the basic unit of classification. The living representative of a genus, a family, or an order is a species. The genus, family, and order are abstract concepts, but the species can be seen and touched.

A species is a group of plants that resemble one another more than they do members of other species. The plants in a species interbreed freely but do not interbreed with members of other species. Although these statements are generally accepted as reliable descriptions of a species, they are oversimplifications because sometimes different species can interbreed to produce hybrids. These hybrids are ordinarily sterile and do not produce offspring, but not always. To complicate matters even more, some species develop seeds without pollination and the subsequent union of male and female sex cells. These seeds germinate and grow into plants that are the exact replicas, or clones, of the parent plant. Every student of botany soon learns that it is difficult to formulate a definition of a species that does not have exceptions. The reader is challenged to explore this topic further in the readings at the end of this book.

Collecting Plants

Humans are collectors of the things that interest them, from bottle caps to vintage cars. It is not unusual, then, that people who are interested in plants should collect plants. It is likely that they are individuals who love the outdoors. Collecting plants not only satisfies their collecting desires but also provides exercise and fresh air. It is a pleasurable activity but one that should be pursued with some caution.

Where to Collect

Plant collectors do not have the freedom to collect all the plants they want wherever they see them. Most of the land surface in the United States is owned by someone or some organization. To avoid legal entanglements, permission should be acquired before collecting on private property. Collecting is prohibited in some municipal, county, state, and federal parks, but limited permission can sometimes be granted if park managers are approached with tact.

Where and What Not to Collect

Even after permission has been granted to collect in a particular area, the collector is not absolved of all responsibility. Out of consideration for environmental conservation, it is a good practice to follow a few simple rules of conduct. When there are only a few plants of a species growing in an area, it is best to collect where they are more abundant. If there is only one plant of that species growing there, it should *never* be collected. The collecting area should be altered as little as possible by the collector. Collecting the last specimen of a species eliminates the colony from that area.

State conservation departments can provide lists of rare and endangered plants for their states. To avoid extinction of these species, every effort should be made to assure their survival and perpetuation in the natural world. They should not be collected. A suggested alternative is to collect their seeds and grow your own. Mature seeds can be harvested without damaging the plant, and it will be a challenge to try to create environmental conditions under which they will germinate and grow.

Tools for Collecting

Experienced collectors usually have a kit that contains the essential tools. It may be stored in a backpack or in the trunk of a car. The kit should be readily accessible at all times because good plant specimens are sometimes spotted when not on specific collecting trips. The basic items that should be included in the kit are a cutting tool, containers for the specimens, notebook and pencil or pen, identification tags, and a hand lens. These are described in more detail below.

Cutting Tool

Every collector needs a cutting tool of some type such as a penknife or a pair of hand clippers or pruning shears. Any kind of pocket knife with a sharp blade is satisfactory, but sometimes for woody plants, hand clippers are better.

Plant Containers

The traditional container used by professional botanists for plant collecting is called a vasculum. It is usually constructed of a light metal such as aluminum, with an easily opened and closed lid, and equipped with a shoulder strap. When the vasculum is lined with wet newspaper, it will keep plants from wilting for several days. These containers can be purchased from biological supply houses but are rather expensive. A selection of scientific supply houses are listed at the end of this chapter.

Plastic bags are less expensive, easier to store and transport, and even many professional botanists are finding them more convenient than vascula. For smaller plants, bags that have a zipper-style closure are very satisfactory. Larger bags that are closed with a twist-tie can be used for larger specimens. Bags of several different sizes, from sandwich size for small plants to very large ones, should be included in the collecting kit. Experience will teach the best mix of sizes to have available.

To keep collected samples from wilting, a piece of wet newspaper can be placed inside and the bag should be kept out of direct sunlight. Specimens prepared in this manner will remain fresh for two days or more. If the plastic bags or vascula are stored in a refrigerator, the specimens will remain in good condition for up to a week. Under no circumstances should plants be frozen if they are to be pressed or dried. When frozen specimens are thawed, they appear to have been cooked.

Notebook

The importance of a field notebook cannot be overemphasized. A record of each species collected should be made on the spot if possible. A collecting trip may yield several species. If recording the data on these is postponed until the end of the day, details may be forgotten or the collection site of one species may be confused with that of another. The data recorded for each plant collected should include the habitat, such as dry sunny hillside, moist shady woods, margin of a cultivated field, edge of a swamp, and so forth. The geographical location should be noted with as much detail as possible. If U.S. Geological Survey topographical maps are available, rural roads, wetlands, fields, and forests will be identified and the latitude and longitude can be determined. For information on USGS topographical maps and how to acquire them, write to:

United States Geological Survey, Map Distribution
1200 Eads Street
Arlington, VA 22202

Knowing the exact location of the site and the date of collection are important if the collector wishes to return at another season for flowers or fruits. The information obtained may also be helpful to other collectors.

Identification Tags

Each specimen collected should be identified with a number or letter. Suitable tags can be inexpensively purchased at almost any store that sells office supplies. The field notebook entry should be listed under this number and a tag should accompany the specimen at all times. It can be attached directly to the plant or placed in a bag with only one specimen in it. The identification number or letter on the tag should be written with a pencil or a pen that does not smudge or smear when moistened.

Hand Lens

An item that may not be essential but can be very useful to the collector is a small hand lens. One that magnifies about ten times is sufficient for most uses. The hand lens is especially useful when examination of flower parts is necessary for the identification of a species.

The Specimen

Herbaceous Plants

When collecting herbaceous plants, it should be kept in mind that the single most important features are the flowers because they are necessary for identification. If the plant is to become part of a collection, it should be in bloom when it is collected. Sometimes identification is easier if both flowers and fruits are available. Some plants bloom over a period of time so that both flowers and fruits can be collected on the same specimen. Usually, though, if fruits are required it will be necessary to return to the collection site later in the season. Since the flowers are essential for identification, if the plant is unknown to the collector it is useful to collect a few extra. This will allow for the dissection that is often necessary for identification, without damaging the specimen for the collection.

The ideal specimen should be one that is representative of the species. It should not be the largest or smallest plant in the colony but rather near the size of most plants of that species at that location. The specimen should be in good physical condition with a minimum of insect damage. There should be enough leaves to clearly demonstrate whether they are attached to the stem in pairs (opposite) or singly (alternate).

Sometimes for smaller plants the entire specimen, including the roots, can be taken. These should be removed carefully so as not to damage or deface the collecting site. When collectors leaves a collection area, it should look exactly the same as before they arrived. For some larger plants, usually only the upper portion of the stem with its leaves and flowers is collected. In addition to a leaf-bearing stem, some plants also have leaves, called basal leaves, that grow directly from the rootstock. Whether they are the same or different from the stem leaves is sometimes an important identifying feature. When plants have basal leaves, a few of these should also be collected.

Ordinarily one specimen of a species is enough for most collectors. If the species is less than abundant, collectors should be guided by good conservation practices and limit the number of samples to one. When a species is plentiful, two complete specimens may be taken in case one becomes damaged. One specimen can sometimes be used for confirmation of identity by sending it to an expert botanist. It is usually not necessary or a good plant conservation practice to take more than two specimens.

Woody Plants

Many trees and shrubs can be identified by leaves alone, but some require fruits or seeds. For example, the color of mature fruit is helpful in distinguishing the species of dogwood. Acorns and mature nuts are essential in the identification of species of oaks and hickories. The specimen for a tree or a shrub should consist of a twig from the end of a branch with enough leaves to show leaf arrangement clearly. Identification almost always requires knowledge of whether the leaves are opposite or alternate. The end of a branch that developed in sunlight is best for this purpose. A twig growing in the shade sometimes grows so slowly that the distance between alternate leaves may be so short that they appear to be in pairs or whorls.

The central part of the stem is called the pith. The appearance of the pith in a twig is sometimes used in identification. To expose this part of the stem use a razor blade or a sharp knife to split the cut end of the twig and excise the upper half of the split section.

Identifying the Plant

If a plant that has been collected is unknown to the collector, it is easier to identify as a fresh specimen than as a pressed and dried one. It is advisable, then, to identify the plant as soon as possible. Most plant manuals and handbooks include dichotomous keys for the identification of unknown species. Dichotomous keys are based on the assumption that any collection of plants can be divided into two groups by an observable characteristic that is present in one group but not in the other.

When comparing two specimens of the same species, there can be much variation in physical characteristics. For example, two plants grown under different environmental conditions may be different in height and thickness of stems and the number, shape, and size of leaves. In these same two plants, though, there will be very little variation in flower structure. In most instances these are observable with the naked eye but sometimes a simple hand lens is helpful.

The use of a dichotomous key can best be illustrated with a small group or plants. Consider a collection with the characteristics listed below.

Plant 1: 10 white petals, 5 stamens, several pistils
Plant 2: 5 white petals, 10 stamens, 1 pistil
Plant 3: 3 white petals, 6 stamens, 1 pistil
Plant 4: 6 blue petals, 6 stamens, several pistils
Plant 5: 5 blue petals, 5 stamens, several pistils
Plant 6: 3 blue petals, 3 stamens, 1 pistil

A dichotomous key for these plants is given below.

A. Plants with petals and stamens in numbers divisible by 4 or 5
   B. Petals blue .... Plant 5
   B. Petals white
      C. Pistil 1 .... Plant 2
      C. Pistils more than 1 .... Plant 1
A. Plants with petals and stamens in numbers divisible by 3
   D. Petals white .... Plant 3
   D. Petals blue
      E. Pistil 1 .... Plant 6
      E. Pistils more than 1 .... Plant 4

In this key, the contrasting statements are given in uppercase letters. The user must repeatedly choose one characteristic over another in progressing to the identity of the unknown plant. Two observations are in order for this type of key.

1. Each of the contrasting statements from which the user must choose are the same number of spaces from the left margin.

2. The contrasting statements often begin with the same word followed by a word or statement that expresses a different condition, for example, "Petals blue" or "Petals white."

There are ways, other than the one above, that dichotomous keys can be organized, but they all require a series of choices between characteristics in arriving at the identity of an unknown plant. The above key is an over-simplification because it involves a very limited group. The keys in plant manuals are much more complex because they cover a greater number of plants.

Books on plant identification are listed in the bibliography.

Preserving the Collection

A herbarium is a collection of pressed, dried, and mounted plant specimens. The objective of collecting plants for most amateur botanists is to accumulate a personal herbarium. This differs from the professional botanist only to the extent that the latter usually collects for an institution such as a college, university, or botanical garden. In order to be useful, a plant specimen must be properly prepared and include essential collecting information. A method that has been successfully used by botanists for many years is to press the specimen flat, let it dry, then attach it with glue to a sheet of white paper. This procedure is the same for both amateurs and professionals. Mounted in this manner, and protected from insects, the specimen will last for hundreds of years.

Equipment Needed for Pressing

Collectors will develop individual routines for preparing specimens, but some basic equipment that will be needed are a plant press, corrugated cardboard ventilators, blotters or newsprint, mounting paper, labels, and an adhesive. These items and their uses, with some alternatives, are explained below.

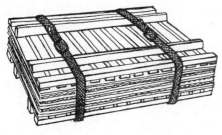

8.1. Plant press

The Plant Press

The function of the plant press is to thoroughly flatten the specimen and hold it that way until it dries. This is accomplished by two solid or wooden grid frames held together by two ropes or straps (fig. 8.1). The plant to be pressed is placed between the two frames, which distribute the pressure evenly. The straps can then be tightened to the desired amount. Plant presses can be purchased at biological supply houses, or one may be constructed inexpensively. Two pieces of quarter-inch plywood or perforated masonite, each twelve by eighteen inches, will serve satisfactorily as frames. Two pieces of heavy cord or, preferably, canvas straps with buckles, each about five feet in length, can be used to hold the frames together.

Corrugated Cardboard Ventilators

Cardboard ventilators can be purchased from biological supply houses or they can be made from corrugated cardboard boxes. Ventilators cut from boxes should be twelve by eighteen inches with the corrugations parallel to the twelve-inch side. These allow air to pass freely through the plant press for rapid drying of the pressed plant.

Blotters or Newsprint

Blotters or newsprint are in direct contact with the plant and absorb juices that may be squeezed from it as it is pressed. If twelve-by-eighteen-inch (30 x 45 cm) blotters are not available, pages of newsprint folded in half are approximately twelve by fourteen inches, and are suitable substitutes. Three or four pages of newsprint folded in half will perform the same function as a blotter.

Mounting Paper

Some collectors may wish to mount their plant collections on the pages of scrapbooks. An advantage of this is the great variety in the types of scrapbooks available and the ease of displaying and viewing the collection. A major disadvantage is that most scrapbook pages are smaller than standard herbarium sheets, which are $11^1/_2$ by $16^1/_2$ inches (27.5 x 40 cm). This is the size of mounting paper used in all professional herbaria. The personal herbarium of the collector will be of greater value if its specimens are compatible with those of professional herbaria. Mounting paper can be purchased at biological supply houses.

Labels

The sheet on which the specimen is mounted must have a label. Commercial mounting paper can be purchased with the label already printed in the lower right-hand corner of the sheet. Plain paper is less expensive, and standardized printed labels can be purchased separately or easily made. The label must provide several items of essential information. Obviously the first item should be the name of the plant. The manual that is used to identify the plant will give the botanical and common names, and the label should carry both. The botanical name should be listed first. Sometimes it will be the only name since some species have no common names. In professional herbaria, the initials of the botanist who named the species are included as part of the botanical name.

The label should also give information in as much detail as possible about the location of the collecting site and the habitat from which the plant was collected. Finally, the name of the collector, the specimen number, and the date the collection was made should be listed.

The specimen number deserves a special mention. Some collectors keep a lifetime list of the plants they have identified or collected and number them consecutively from number one onward. Others prefer to start their numbering anew each year and designate the year of collection as 02–1, 02–2 then 03–1, 03–2 and so on. The specimen number, environmental data, and site location will be provided by the field notebook.

All of this information can be recorded on a label about the size of a three-by-five-inch card (7.5 x 12.5 cm). If you make your own, four labels can be typed on a sheet of $8^1/_2$-by-11-inch (21.3-by-27.5 cm) paper. The following is suggested as a model.

HERBARIUM OF JANE DOE

Botanical Name _____

Common Name _____

Family _____

Locality_____

Habitat _____

Collector _____

Date _____ No. _____

Adhesive

The function of the adhesive is to attach the specimen to the mounting paper. White glue such as Elmer's glue is probably the most convenient for the individual collector. It is readily available from many stores, is very effective, and is used by many professional botanists. Some collectors prefer thin strips of an adhesive linen tape to attach the specimen. This type of adhesive is available at most office supply stores. Transparent plastic tape is unsatisfactory because it dries and yellows with age.

Pressing the Specimen

Some collectors carry a plant press into the field and press the specimens as soon as they are collected. Others prefer to transport the specimens to home base where conveniences such as work tables may be available. Regardless of the location, there is a recommended routine for the process as described below.

1. The bottom frame of the press should be placed on the ground or on a table.

2. A corrugated cardboard ventilator is the placed on the frame.

3. This is followed by a blotter, or in the absence of blotters, several pages of newsprint folded in half.

4. The plant specimen to be pressed is placed on half a page of folded newsprint. It should be spread carefully so that flowers are unobstructed and

there is a minimum overlapping of leaves. One or two leaves should be turned over with the bottom side up, since features of the leaf undersides are sometimes important for identification. If a specimen is too large to fit easily on half a page of newsprint, it can be bent to form a V, or if still larger, bent again to form an N. Then the other half of the newsprint page is folded over the specimen. The name of the plant or its number is written on the outside of the folded newsprint.

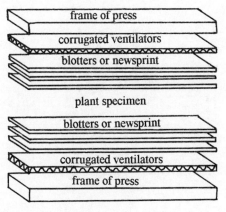

8.2. Construction of a plant press

5. A blotter, or two or three pages of folded newsprint are then placed on top of the newsprint containing the plant.

6. Another corrugated cardboard ventilator is placed on the blotter or newsprint.

7. The process can now be repeated for other specimens in the same order: ventilator, blotter, specimen, blotter, ventilator (fig. 8.2).

8. When all specimens have been so prepared, the top frame of the press is placed on the stack and the straps tightened around each end. Apply as much pressure as possible in tightening the straps. Having someone stand on the press while tightening is helpful.

Drying

The faster the specimen dries, the less likelihood that there will be discoloration of the flowers and leaves. Pressing a plant in a book is not recommended because the specimen dries slowly with practically no air circulation, usually resulting in discoloration of not only the plant, but the pages of the book as well. In a plant press, depending on the temperature, humidity, and the size of the specimen, the plant will dry in five to ten days. After the first twenty-four hours, it can be examined to rearrange flowers or to smooth wrinkles.

8.3. Drying a plant specimen

If faster drying is desired, the plant press can be positioned over a mild source of heat so that warm air rises through the channels provided by the corrugations of the ventilators. The most convenient source of heat is probably an ordinary light bulb. The press can be placed between two chairs with the bulb at least one foot below the corrugations (fig. 8.3). Only mild heat is recommended because overheating may cause the specimens to turn brown. The plants will dry in two or three days with this arrangement.

If blotters and ventilators are not available, the plant press can still be useful. On the bottom frame, place a stack of three or four pages of folded newsprint. On top of these, place the folded page holding the plant to be pressed. Add another stack of newsprint similar to the first. At this point, other specimens can be added following the same procedure. The top frame of the press can now be applied and the straps tightened. It may take a little longer for the plants to dry by this method, but the porosity of the newsprint will provide enough aeration to prevent discoloration. To hasten the drying process, the newsprint can be changed after the first twenty-four hours.

Mounting

When the plant is removed from the press, it is ready to be attached to a sheet of paper. Using white glue that can be squeezed through a nozzle from a tube or other container, the collector can apply dots of glue to several places on the underside of the specimen. After positioning it on the paper, dots of glue can be applied to other points as needed. Sometimes a thin string of glue, when it dries across a leaf or other delicate part, will effectively pin it to the paper. An advantage of using glue is that usually it will last as long as the paper or the plant. A disadvantage is that the plant can never be removed from the sheet.

An alternate method of mounting is to use linen gummed tape. Thin strips of tape can be placed across stems and leaves at critical points to hold the plant on the paper. This method has the advantage of allowing the removal of the specimen from the paper at some future date. A disadvantage is that over a long period of time the tape may dry and lose it adhesiveness.

Attaching the Label

Attaching a label, complete with the name or identifying number of the plant, must be a part of the mounting routine. The most convenient ones are those that are already printed on some grades of commercial herbarium paper. However, gummed labels can be purchased. If you make your own, they can be attached with the same glue that was used to attach the specimen.

Protecting and Storing the Collection

There are two major threats to any herbarium: fungi and insects. Preventing contact of the specimen with moisture is the key to controlling fungal growth. If the mounts are dry at all times and stored in an area that has a consistently low humidity, the threat of fungal attack is greatly reduced. A greater problem is often caused by insects. Even if the plants are completely dry, an infestation may occur. Among the most damaging of the insect pests are several species collectively called dermestid beetles. They are very small beetles that in the larval stages feed on dry plant tissue. The collection should be inspected at least three times a year for indications of fungal or insect damage.

Professional herbaria store their collections in airtight metal cabinets. These can be purchased from biological supply houses, but they are very expensive and probably impractical for the individual collector. Professional herbaria also use large manila folders, called species covers, to hold all the specimens of each species. While these are convenient, they are not essential, and, in their stead, the collector can use folded newsprint pages. Any appropriately sized cabinet or even cardboard boxes will serve as storage facilities. They can be made approximately airtight by splitting large plastic trash bags and tacking or gluing them in as liners. If the collection is mounted in scrapbooks, these can be stored in large plastic bags. It is worth repeating that whatever the storage facility, the storage area should be well ventilated and dry.

There are several types of fumigants that can be used to protect the herbarium from insects. The easiest to acquire is probably paradichloro-benzene, or PDB, which can commonly be purchased as either moth crystals or moth balls. If the storage cabinets or boxes have reasonably tight closure, a small cloth bag of crystals or perforated bags of moth balls can be placed in each compartment. Like scrapbooks, smaller collections mounted on individual sheets can be stored in large plastic bags into which crystals or moth balls have been placed. The chemicals should be renewed about every four months. If the herbarium is stored in the home, it should be placed in an area where family members will not be constantly exposed to PDB fumes.

Some botanists have suggested an alternate method of protecting the plant collection from infestation. Placing the mounts into a freezer for twelve to fourteen days seems to be enough to kill insect pests. This has great appeal to many people because it eliminates the use of chemicals. A disadvantage may be that it requires the periodic availability of a considerable amount of freezer space.

Displaying the Collection

There are several ways that plant mounts may be prepared for display. Collectors often give presentations for school groups, scout groups, 4-H clubs, or other organizations for young people. It may be desirable in these presentations to have specimens that can be handled by the audience. Young, eager, and curious hands can do a lot of damage to a dry and very brittle mounted plant. For collections mounted in scrapbooks, the best books are those having individual pages that are removable and have plastic covers. These provide a measure of protection for the plant and are excellent for viewing.

Collectors who do not use scrapbooks may wish to laminate with plastic the mounts of the specimens they will use for a presentation. However, the cost of this option may be prohibitive. Perhaps a more realistic plan is to attach the plant mount to a standard corrugated cardboard ventilator or other stiff cardboard to prevent bending the specimen, and then wrap it tightly with adhesive plastic kitchen wrap. Plants prepared in this way are suitable for hands-on presentations to groups of all ages.

Sometimes special mounting is appropriate for specimens that are bulky or unusually attractive. For mounts that are flat, the whole sheet can be enclosed in a frame called a botanical mount. It consists of a stiff cardboard back with a glass front held together usually by black tape around the edges. The botanical mount may contain a thin layer of cotton to hold the mount in place. For bulky specimens such as those with thick stems, pine cones, or hard fruits, a type of frame known as a Riker mount is available. This is a shallow cotton-filled box with a pane of glass on one side. The specimen is usually not mounted on a sheet of paper but is embedded and held in place by the cotton. Both of these types of mounts are expensive, but plants mounted in these ways are often so attractive that they can be displayed as wall hangings.

Special Plant Groups

Some collectors may wish to include examples of all major plant groups in their collections. The discussion in the preceding pages has been concerned mainly with methods of collecting and preserving seed plants. These methods are valid for most plants, but some groups require a different type of treatment. The life histories and growth habits of the plant groups listed below are described in chapter 2.

Fungi

Slime molds. The best time to collect this group is when they are in the fruiting or spore-producing stage. The type of sporangium of a species is an important feature for identification. A small section of the rotting wood or soil on which the sporangia are growing should be collected. Since the fragile sporangia may be dry and brittle, care should be exercised in taking the sample. It can be stored in a small box with some moth crystals and with a label on the outside. Boxes of various sizes are available from suppliers, but small boxes around the home, such as those that jewelry comes in, are just as good.

Mushrooms and other soft fungi. As with slime molds, the structures that are collected are fruiting or spore-bearing stages. They can be preserved by drying or in a liquid preservative. For storage in liquid, they should be cut near the ground and placed in the preservative immediately. In the absence of a commercial preservative, rubbing alcohol is satisfactory for this purpose.

If the soft fungi are to be preserved by drying, they should be dried as quickly as possible because they decay rapidly when moist. Applying some form of mild artificial heat will hasten the process and help prevent the onset of decay. After drying is complete, the specimens can be stored in labeled boxes of the appropriate size. Moth crystals should be included as protection from insect attacks. To avoid decay, it is especially important to keep these specimens dry.

Shelf fungi. Some of the shelf fungi are hard and woody. The collector needs only to break them from the log or stump on which they are growing. They may grow to a very large size, but collectors can choose the size that is best suited for their collections. When the fungi are thoroughly air dried, they can be stored in boxes or plastic bags with PDB crystals.

Lichens. A good plan for collecting lichens is to use a plastic bag in the field, then transfer them to boxes or envelopes for storage. The specimen should include the small cuplike spore-producing structures of the fungal portion of the lichen. This is very important for identification. Care should be taken in their transport and storage because they are usually dry and brittle and easy to shatter. In the fungi, as in all plant groups, detailed field notes should be made so that a complete label can be attached to each specimen.

Mosses and Liverworts

These are so small that any sample will contain several plants. As with lichens, they can be collected in the field with plastic bags. Complete specimens should have the spore-producing structures. In mosses, this includes the green bottom portion with the stalk and capsule at the top. After allowing the specimens to air dry for two or three days, they can be stored with PDB crystals in appropriately labeled boxes or envelopes.

Ferns and Fern Allies

Ferns. Ferns, clubmosses, and horsetails can be pressed and mounted in the same way as seed plants. Some special notes on collecting will be useful. In all three groups it is important that the specimen have spore-bearing structures. In many fern species, these are on the underside of the frond. In those that have dissected leaves, they are on the undersides of the leaflets. When pressing the fern leaf, be sure to turn a few leaflets over so the fruit dots, or sori, can be seen when the leaf is mounted. In other fern species, the spore-bearing structures are on separate stalks. These stalks must be included for complete specimens.

Clubmosses. The clubmosses may have spore-bearing structures, or sporangia, in the axils of upper leaves or in cones at the tips of branches. The upright branches of some species arise from a horizontal stem that runs along the surface of the ground or just beneath the surface. The characteristics of this stem are sometimes important for identification, so a small section should be included with the specimen.

Horsetails. The main factor to keep in mind when collecting horsetails is that some species have a spore-bearing, or fertile, stem that appears early and then disappears before the green vegetative shoot is fully developed.

Leaves, Leaf Skeletons, and Leaf Prints

Leaves

A collection of tree leaves is easy to make, and it can be especially helpful in hands-on presentations to youth groups. Fully developed leaves with no insect damage should be selected and pressed for mounting in the manner described for seed plants. They can be mounted one or more leaves per page, as the collector wishes. Leaf collections of woody plants are more useful because these plants can more reliably be identified by leaf characteristics than can herbaceous plants.

An alternate way to make a collection of leaves is to coat them with wax. After they are thoroughly dry, they can be pressed between layers of wax paper with a warm iron. This will apply a thin layer of wax to each side of the leaf. Prepared this way, the leaves can be either mounted on paper or stored in a box or plastic bag and they will last for years. Leaves that are at the peak of autumnal coloration can be waxed in this manner and, although the color may fade after a few weeks, they can be used for attractive seasonal decorations.

Leaf Skeletons

Another way of displaying a leaf collection is by making leaf skeletons. Leaves are made up of a network of tiny veins. When the interconnecting tissue is removed, a beautiful lacy outline of the leaf remains. There is no quick or easy way to make a leaf skeleton, but when successfully done, the result justifies the effort.

There are chemical methods that can be used for skeletonizing leaves, but they require chemicals that may not be readily available to the individual collector. The easiest method is to let nature take its course. In a gallon of water, add two tablespoons of forest soil humus or other rich topsoil, to insure the presence of decay bacteria, and two tablespoons of sugar or heavy syrup to stimulate bacterial growth. Immerse leaves in the solution and let it stand for a month in a warm place. Then remove a leaf and wash it with a gentle stream of water to remove the soft tissue. If a month is not enough time for complete leaf decay, let the solution stand for another two or three weeks. The leaf skeletons can be mounted between glass plates for projection on a screen, or they can be attractively displayed in a scrapbook or on mounting sheets.

Leaf Prints

A different way of making a leaf collection is to make a collection of leaf prints. This involves the use of printer's ink, which can be acquired at print shops or office supply stores that sell ink for rubber stamp pads. Several methods have been recommended but one that is simple and produces good results is described below.

With a small soft brush, such as a pastry brush, apply printer's ink to a piece of masonite or other hard, smooth surface. Place the leaf, underside down, on the inked surface. Cover it with a sheet of newsprint and roll it very gently with only the weight of a rolling pin or a large drinking glass. The objective is to bring all the veins on the underside of the leaf into contact with the ink. Remove the leaf from the inked surface and carefully place the inked side on a sheet of white paper. Cover it with newsprint and again roll it gently, making sure it does not move. The quality of the print will be determined by the amount of ink picked up by the underside of the leaf. With practice, this can be controlled by the amount of pressure applied when the leaf is on the inked surface. Prints with varying degrees of density can be made according to the wishes of the collector (fig. 8.4).

An interesting project with leaf prints is to show the leaf variations of a given species. Some trees have noticeable leaf variations, such as those of sassafras *(Sassafras albidum)* and mulberry *(Morus rubra)*. Smaller but observable variations can be seen in the leaves of other species such as white oak *(Quercus alba)* and sugar maple *(Acer saccharum)*. These can be used to demonstrate the influence of genetic or environmental variation within a species.

8.4. Leaf print

Photographing Plants

A camera can be a great asset for the plant collector. Almost any type of camera will suffice, but one that allows close-up focusing is recommended. It is also useful sometimes to have a camera with film that can be processed into slides for projection on a screen. Pictures of the collecting sites can also be taken. These provide an added measure of authenticity when attached to mounted specimens. In addition, a picture of a plant as it grows in the wild is often helpful in identification.

Some individuals confine their collecting to what can be captured on film. This usually results in large numbers of color slides of wildflowers. As these accumulate, a system of organization for storage becomes a necessity. They may be organized by habitat, such as plants of wetlands or forests; or by geography, such as plants of New York or West Virginia. As familiarity with botanical classification increases, the collector may want to organize the slide collection by plant family, such as plants of the lily family or aster family. A collection of plant slides, organized in any way the collector chooses, is rewarding for private or public showings. Specially constructed boxes for slide storage can be purchased from supply houses.

The camera can also be used for time-lapse photography, which can yield spectacular results. Taking daily photographs of a germinating seed or hourly photographs of a flower as it opens are exciting activities. Setting a camera tripod in exactly the same location for color photographs of collecting sites in each ecological season will provide valuable life history information. For information on the ecological seasons and photographing them, see chapter 6.

Drying Plants Without a Plant Press

Dried flowers are often used in wreaths, swags, and other home decorations. These are obviously not dried in plant presses; there are alternate ways of drying when the objective is a three-dimensional rather than a flat specimen. The simplest method is to collect the plants at their flowering or fruiting peak, tie them in bundles, and hang them upside down in a dry, dust-free, protected place. Plants with many small flowers clustered in dense heads lend themselves to this kind of drying. A few examples of this type of plant are yarrow *(Achillea millefolium)*, baby's breath *(Gypsophila paniculata)*, Joe-Pye weed *(Eupatorium spp.)*, forget-me-not *(Myosotis spp.)*, purple coneflower *(Echinacea purpurea)*, goldenrod *(Solidago spp.)*, and Oswego tea *(Monarda didyma)*.

A method that has long been used to dry flowers is to bury them in sand. To use this method, cover the bottom of a container with one or two inches of clean, sifted beach sand. Place the flowers on this layer, stems up, and very carefully cover them, making sure the sand is between and around each petal and delicate part, to hold them in their natural positions. Cover the flowers with one or two inches of sand and store the container in a dry place for about two weeks. Then very carefully pour off some of the sand to see if the petals are stiff and dry. If they are not, more drying time will be necessary.

Sand is not a drying agent. It serves as a frame to hold the buried flowers while they dry naturally. Other substances that have been suggested to serve this function are cornmeal, diatomaceous earth, powdered pumice, and even dry cereals such as cream of wheat. Very often these substances are mixed with an active dehydrating agent such as borax. Borax and sand, borax and cornmeal, powdered pumice and cornmeal, and pure uniodized salt have all been recommended for drying flowers.

A commercial dehydrating agent called silica gel is widely used for more rapid drying. It has a sandlike consistency and usually contains crystals of cobalt chloride. These are indicator crystals that are blue when the substance is dry but turn pink when the silica gel has absorbed all the water it can hold. An advantage of using this agent is that it can be dried in a regular or microwave oven and used again and again. In the drying process, when the crystals return to their blue color, the drying agent is ready for reuse.

The method of drying flowers with silica gel is the same as that suggested for sand except it should be in a container with a lid. The container should be closed during the drying process, which may take two to seven days, depending on the number and size of the flowers being dried. For microwave drying the time is much less. Silica gel should be handled with care and kept out of the reach of children because the dust can cause irritation to respiratory tissues. Silica gel is available from drug stores and craft stores, and the container usually has detailed directions for its use.

Scientific Supply Houses

A few supply houses are listed below where the equipment described in this chapter may be purchased.

Carolina Biological Supply Company
2700 York Road
Burlington, NC 27215

Central Scientific Company
3300 Cenco Parkway
Franklin Park, IL 60131

Frey Scientific
905 Hickory Lane
P.O. Box 8101
Mansfield, OH 44901

Wards
P.O. Box 92912
Rochester, NY 14692

Sargent-Welch Scientific
911 Commerce Court
Buffalo Grove, IL 60089

9

Activities and Investigations

1. Determining the Age of Trees

A. Observing the Age of Twigs

Examine the tips of branches on a tree in your yard, in the park, or in a woodland. This can be done in any season.

In winter. Locate the bud at the tip of the twig. Observe the bud and note that it is covered by overlapping plates (a small hand lens will be helpful). These are bud scales, and each one is attached to the stem at the base of the bud. When the bud begins growth in spring, the bud scales are shed leaving a ring of scars where they were attached.

Farther back on the twig, a ring marking the location of last winter's bud can be seen. The space from the ring of bud scale scars to the end or terminal bud is the extent of last year's growth of the twig. If the twig was in sunlight, this distance may be one or two feet in length. If it was in shade, the distance may be very short. Farther back other rings can be seen that mark the growth of the twig in previous years.

In summer. The first ring of bud scale scars marks the beginning of the current season's growth.

The bud scale scars are eventually obscured by the formation of bark, but sometimes five or six years of growth can be identified on a twig. If your observations are being made in winter, can you guess whether the twig was growing in sun or shade?

B. Counting Annual Rings

The most reliable method for determining the longevity of a tree of un-known age is counting its annual rings. Professional foresters and den-drochronologists use an instrument known as a Swedish increment borer. This is a hand-operated drill with a hollow bit that takes a slender core of wood from the bark to the center of a tree. The drill leaves a very small hole in the tree that closes soon after the drill is removed, causing no damage to the tree. Most people do not have access to Swedish increment borers, but annual rings can be observed on the stump of a recently cut tree or even a piece of firewood. The beautiful patterns on stained or varnished wood fur-niture or fixtures are the result of the angle at which the rings were cut. The rings can be recognized in these patterns.

C. Measuring the Percentage of Increase in Size of One Year's Growth

Before the buds begin to swell in spring, measure the circumference of a tree about four feet above the ground. Use a metric tape and measure to the nearest millimeter. Mark the exact location and measure again when the leaves begin to show autumnal coloration. To get the percent of increase in size of one year's growth divide the amount of increase in circumference by the spring measurement.

An interesting long-term project is to take these measurements for sev-eral consecutive years and keep records of rainfall during the growing sea-son. This data will probably show a relationship between the amount of rainfall and the amount of increase in circumference. Usually trees grow more during wet seasons than during drier ones.

D. Calculating the Age of a Tree by Measuring the Yearly Increase
   in the Circumference of Its Trunk

Determine the diameter of the tree you selected by the formula for the cir-cumference of a circle ($C = \pi D$: circumference equals pi (π) times the diam-eter; pi is always $^{22}/_{7}$ or 3.14). Assume the tree you have measured has a circumference of 96 centimeters.

$$C = \pi D$$
$$96 = 3.14D$$

By dividing both sides of the equation by 3.14 it becomes:

$$D = \frac{96}{3.14}$$

$$D = 30.5$$

Therefore a tree with a circumference of 96 centimeters has a diameter of 30.5 centimeters.

Suppose your tree has a circumference of 98.5 centimeters when you measure it in autumn. Using the above formula, it can be determined to have a diameter of 31.3 centimeters. This is an increase of 0.8 centimeters in one year of growth. Assuming a constant rate of growth throughout its life, this tree is about thirty-nine years old ($31.3 \div 0.8$). Using the same assumption, in ten years the tree will have a diameter of 39.3 centimeters ([10 X 0.8] + 31.3) and a circumference of 123.4 centimeters (3.l4 X 39.3).

This method of determining the age of a tree gives only a rough estimate, at best, because trees do not usually grow at the exact same rates each year. Variation in any one of several factors may influence the growth rate. In addition to responding to the amount of rainfall, tree growth may also be influenced by changes in exposure to sunlight. The removal of structures or other trees that block the sun may increase the growth rate. Any factor that increases or decreases the availability of mineral nutrients in the soil will increase or decrease the growth rate. For example, applying fertilizer to a lawn may increase the growth rates of trees growing there.

2. Does Moss Grow on the North Side of a Tree Trunk?

In fiction, an individual lost in the forest frequently finds the way out by observing that moss grows on the north sides of tree trunks. Is this the truth or myth? Three types of growth may be seen on tree trunks: (1) green algae will appear as a green layer that does not rise above the surface of the bark, (2) mosses are small green plants with tiny stemlike and leaflike structures, and (3) foliose lichens, like mosses, stand above the surface of the bark (see chapter 2). After the following investigation, decide for yourself whether or not these organisms, or any combination of them, grow on the north sides of tree trunks.

Using a pane of glass or a piece of transparent plastic, outline a square of at least fifteen centimeters (cm) on a side and mark off a grid of square centimeters. A square with 15 cm on a side will have and area of 225 square centimeters (15 X 15 = 225). With a compass locate the north, south, east, and west sides of a tree trunk. For ease of measurement, the tree should be at least 30 centimeters (1 ft.) in diameter at breast height. Place the grid in the center of the trunk on each of the four sides at 30 centimeters above the ground. Estimate the total number of square centimeters covered by lichens, algae, and moss. For example, if there is growth covering 50 square centimeters on one side of the trunk, this can be expressed as a percentage (50 ÷ 225 X 100 = 22.2 percent). Repeat this process at 60, 90, and 120 centimeters (2, 3, and 4 ft.) above the ground. Perform these measurements on several trees. The four sides can then be compared for the greatest percentage of plant cover. Now, if you were lost in the woods, would the growth on tree trunks help you find your way?

As a refinement to this investigation, place the grid on the northeast, northwest, southeast, and southwest sides of the tree trunks and determine the percentage of coverage at the same levels as before. Do these results change your conclusions? You can also determine the relative abundance of lichens, algae, and mosses. Is one of these more common than the others on any side of the tree trunks? Is the growth about the same at each of the levels above the ground?

3. Dormancy

In the northern portions of the temperate zone, many species of plants become dormant during the winter months. For these species, dormancy is more than simply a cessation of growth because of low temperatures. It is a chemically induced state that requires a period of chilling before continuation of growth. This has a profound influence on the plant's geographic distribution. If a plant with a chilling requirement is not exposed to the required minimum cold period, it may not break dormancy in spring; or if it does break dormancy, growth may be weak and the plant devoid of flowers. Thus a species may be limited in its southward distribution by length and severity of winter temperatures. One of the reasons that apple trees do not flourish in Florida is because winters there are not cold enough or long enough to fulfill the dormancy requirements of apple trees.

Bud Dormancy

The marker of ecological spring in the deciduous forest is the swelling of buds (see chapter 6). This indicates they have broken dormancy. The event is preceded by increasing temperatures and day lengths. The required period of chilling may have been satisfied weeks or months earlier. The durations of chilling and warming are not known for most species of forest trees. Those that grow in northern regions may require longer periods than the same species in more southerly regions. It can be assumed that after the chilling requirement has been achieved, the buds will resume growth if temperature and light conditions are favorable.

The beginning of ecological winter is indicated by the fall of leaves. By the time this occurs the trees are in a state of dormancy. They will remain dormant at least until the required period of chilling has been achieved. The length of the period can be investigated in a superficial way by using twigs. When twigs are cut from the ends of branches and brought inside, they should be cut again underwater so that air bubbles do not plug the ends of water-conducting cells. The cut end should be kept in water, at room temperature, under twelve to fourteen hours of light per day. The twig should be inspected daily for at least a month. The first sight of any structure emerging from the bud signals the breaking of dormancy and the resumption of growth. If there is no change in a month, it is probably an indication that the chilling requirement has not been satisfied.

The first twig should be collected immediately after the fall of leaves. In some regions, leaves are on the ground by November 1. If this is the date of the first collection, others should be made, following the same procedure, on December 1, January 1, February 1, March 1, and April 1. If the buds resume growth on twigs collected on one of these dates, the chilling requirement has been satisfied by that date for that species.

This method can be used to force early flowering of some ornamental species. By cutting twigs in late winter, pussy willow, azalea, lilac, and forsythia can sometimes be brought to flower long before they would bloom naturally.

4. Investigating Plants Using Pollen, Spores, and Gametophytes

A. Observing Moss Spore Dispersal and Germination

When moss spores germinate, they form green algal-like filaments called protonemas. Individual filaments are very difficult to see, but they may occur in networks made up of many strands. A diligent observer may be able to see them as green threadlike lines on moist soil usually near moss plants (see chapter 2).

The mechanism for the dispersal of moss spores may be seen with a simple hand lens. An intact moss plant complete with gametophyte and sporophyte can be collected and taken indoors. The capsule of the sporophyte has a cap that covers a ring of enfolded flaplike teeth. If the moss plant is placed under an incandescent lamp, the capsule will dry, the cap will pop off, and the teeth will flip outward expelling the spores. When the capsule is remoistened, the teeth will return to their enfolded position.

B. Cultivating Fern Gametophytes

Fern gametophytes commonly grow in areas where fern plants (fern sporophytes) are abundant, but they are difficult to find because they are so tiny. They are fairly easy to cultivate. Thoroughly clean a three or four inch clay flower pot and boil it for ten minutes, then pack it tightly with moist peat moss or shredded paper. Invert it in a dish of water so that the water is in contact with the contents of the pot. Add water to the dish as needed. Cover the dish of water and its inverted flower pot with a transparent glass or plastic bowl to keep out dust and fungal spores.

Fern spores are easily collected from sensitive and ostrich ferns, both of which have separate specialized spore-bearing leaves. Sensitive fern is very common in damp areas. By gently tapping the spore-bearing leaves on a white sheet of paper, the brown spores can be collected. Uncover the inverted flower pot and sprinkle some of the spores on its surface. A common error is to dust the spores too heavily; try using a medicine dropper and spread them very sparsely.

Place the covered pot where it is out of direct sunlight and mature fern gametophytes will develop in eight to ten weeks. Sporophytes may be visible in about eight weeks but will usually be plentiful in six months. The young fern plants can be transplanted in moist soil after their roots have developed.

C. Observing and Germinating Horsetail Spores

Horsetail spores are produced in cones at the ends of stems. Each spore has four long slender appendages, the elaters, that are wrapped around it as it develops. When the cone matures and the spores dry, the appendages extend like four twisted helicopter blades. This expansion by all spores ruptures the sporangium, resulting in the spores' release. The expanded appendages serve as wings that aid in dispersal of the spores by air currents. A spore with its extended appendages can be seen at the lower range of vision with a hand lens of 15X or 20X magnification. When the spores are moistened, the elaters contract and wrap around the spore.

If the spores are sown on the surface of water in a bowl, they will germinate within a few days but will not develop to completion.

D. Cultivating Horsetail Gametophytes

A complete horsetail gametophyte can be grown if newly formed spores are sown on a layer of clean, moist peat moss that has been boiled to kill fungal spores. The peat moss should be placed in a dish, sown lightly with spores, covered, and placed out of direct sunlight. If the peat moss is kept moist, gametophytes will develop in a few weeks.

The gametophyte looks like a tiny green pin cushion about the size of a pinhead or ranging in size from one millimeter to one centimeter in diameter. The gametophyte produces both egg and sperm cells, and several horsetail plants (sporophytes) may arise from the same gametophyte if more than one egg cell is fertilized.

E. Growing Horsetails from Cuttings

Cuttings from both field horsetail and scouring rush will root if planted in wet sand. The cuttings should be embedded in the sand to a depth that includes at least one joint of the stem. The stems should root in a week or so and will continue to grow if the sand is watered regularly.

F. Germinating Pollen

The pollen grain of seed plants is the immature male gametophyte, and it produces the sperm nucleus. In most seed plants, the sperm is nonmotile and is delivered to the vicinity of the egg cell by a pollen tube. In flowering plants, the pollen grain reaches the stigma by wind or an animal pollinator. It germinates there and grows through the style to the ovules in the ovary. After the egg cell in the ovule has been fertilized by the sperm nucleus carried in the pollen tube, the ovule becomes a seed. One pollen grain is thus required for the development of each seed produced. Many seed plants produce great quantities of pollen, especially those that are wind-pollinated such as birch, oak, and pine.

In many species, the pollen grain will germinate and the pollen tube will begin growing if the grain lands on water. To observe this fill a shallow bowl or pan with water and sprinkle some freshly collected pollen on the surface. In a day or two, some grains with beginning pollen tubes may be seen with a 14X or 20X magnifying hand lens.

G. Making Spore Prints of Mushrooms

Spore prints are easy to make and they can be useful in the identification of mushrooms. For instructions, see chapter 2.

5. Investigating Plants by Sample Plots

In previous chapters, habitats have been discussed with lists of plant species that grow there. This descriptive information is very important, but for some types of information quantitative mathematical measurements are necessary. For example, in order to understand a forest ecosystem it is important to know how many species make up the canopy, what their sizes are, how they are spaced, and which ones exert the most influence on the forest floor. The most accurate way to acquire this information would be to count, measure, and identify all the trees. Counting every tree in a forest is not practical, so ecologists have developed methods of estimating composition, spacing, and cover. Two of these methods are described below.

A. The Quadrat Method

By this method, sample plots that are of a manageable size, called quadrats, are marked. In each quadrat, all the plants can be identified and measured to give estimates that are valid for the entire woodland, meadow, or wetland. The size of the quadrat is of great importance because it must be large enough to give a representative sample of the vegetation. A method that has been used to determine the appropriate size is an arrangement called nested quadrats (see fig. 9.1). In this procedure, progressively larger quadrats are sampled and additional species added to those from all previous plots. When a point is reached where the additional larger plot yields no more than 5 percent of the total number, the previous plot containing 95 percent of the species can be accepted as the appropriate sample size.

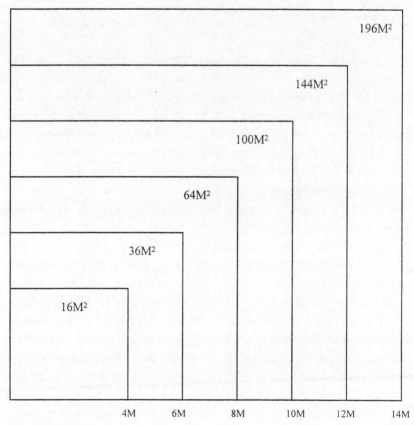

9.1. Nested Quadrats

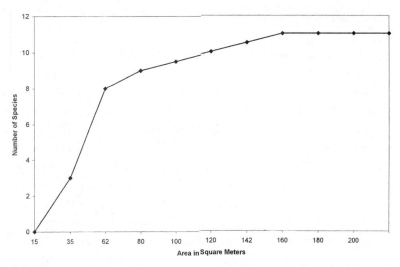

9.2. Species Area Curve

Another way to determine the size of the sample quadrat is by graphing a species area curve. This can be accompanied by plotting the number of species along one axis and the area of the quadrat along the other (see fig. 9.2). The curve will rise sharply at first, then begin to level off as the size of the quadrat increases. As it begins to level off, this marks the minimal size of the area to be sampled. The point on the graph just before the curve becomes horizontal marks the appropriate size of the sample plot. In figure 9.1, the quadrat is 10 meters on each side, or 100 square meters. Since the metric system is used in scientific measurements, meters are the units of measurement in figures 9.1 and 9.2.

The quadrat method of sampling is often used to gain information about the canopy and secondary tree layers (see chapter 5). The minimum size of trees to be counted should be decided at the outset. A size that is sometimes used is ten centimeters (4 in.) in diameter at breast height (usually abbreviated as DBH, about $1^1/_3$ meters or $4^1/_2$ feet above the ground). By counting smaller trees and seedlings, information can be obtained on relative rates of reproduction. If the proportion of species among the seedlings is different from that of mature trees, it suggests that the composition of the forest may be changing.

When a suitable size for the quadrat has been selected and the trees within it have been identified, counted, and measured, several characteristics of the forest can be determined. Three of these are density, abundance, and cover.

Density. Density (D) is the number of trees per unit area. If the size of the quadrat is one hundred square meters (100 m²) and the total number of trees is twenty, the density is twenty trees per hundred square meters. It can also be expressed in the following way:

$$D = \frac{\text{total number of trees}}{\text{total area of quadrat}} = \frac{20}{100} = .2 \text{ of a tree per m}^2$$
$$\text{or one tree per 5 m}^2$$

Abundance. (Ab.) Abundance is the number of individuals of a species expressed as a percentage of the total number of individuals of all species. For example, suppose that of the twenty trees in this quadrat, six are sugar maples. Thirty percent of the trees in this forest are sugar maples.

$$\text{Ab.} = \frac{\text{total number of individuals of a species}}{\text{total number of individuals of all species}} = \frac{6}{20} \times 100 = 30\%$$

Cover. Relative cover (R.Cv.) is the percentage of quadrat area under the canopy of a given species. To estimate this, the area of the cross section of each tree (basal area) should be calculated. It is assumed the trees with the greatest basal areas have the largest canopies and thus the greatest coverage. To determine basal area, the diameter at breast height of each tree should be recorded. A meter stick can be used to estimate this measurement by line of sight. Then using the formula for the area of a circle, $A = \pi r^2$, the basal area can be computed where A is the area, π is 3.14, and r is one-half the diameter. In a forest, the total basal areas of all trees in the quadrat can be accepted as 100 percent coverage. The relative basal area of each species can be expressed as a percentage of the total of all trees.

Suppose the total basal area for all trees in the quadrat is 15,000 square centimeters and the total for sugar maples is 4,300 square centimeters.

$$\text{R.Cv.} = \frac{\text{total basal area of sugar maple}}{\text{total basal area of all species}} = \frac{4,300}{15,000} \times 100 = 28.6\%$$

Sugar maple covers 28.6 percent of this woodland. Relative cover is sometimes called relative dominance. It indicates the degree of dominance of each species in the sampled woodland.

The quadrat method of sampling trees can also be used for shrubs and herbaceous plants. Usually for smaller plants, the size of the quadrat is smaller. If a sample plot ten meters on a side is used for trees, one with four meters on a side is satisfactory for shrubs, and one meter on a side for herbaceous plants. Although this method of acquiring information about plant communities is time consuming, it provides results that are reasonably reliable.

B. Plotless Methods

These are methods most often used for sampling trees rather than shrubs or herbaceous plants. They are based mainly on the distances between ran-

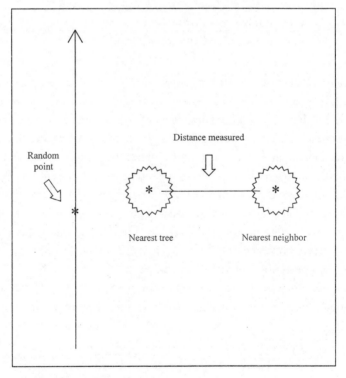

9.3. Nearest Neighbor Method

domly selected trees in a woodland. The information obtained by plotless methods is not as reliable as that collected by the quadrat method, but it takes less time to acquire.

Nearest neighbor method. To use this system of sampling, an imaginary line is established through a forest. It can be a line of sight or one maintained with a compass. The line can be straight throughout or it can include sharp turns to keep the sample area at a reasonable size. If the line is to include turns, their spacing should be established before sampling begins.

A point is designated along the line and the tree nearest to it is identified. The tree nearest to the first tree is identified and the distance between them is measured (see fig. 9.3). In order to compute cover, the DBH should be taken for each tree. The next point be selected along the line should be far enough from the previous point that neither of the two previously identified trees are included in the next pair.

This process should be repeated until a designated number of tree pairs have been identified. Suppose fifty pairs of trees have been recorded and the average distance between them is five meters. The density per hectare (10,000 m^2) of this woodland can be estimated by the following formula.

$$D = \frac{10,000}{1.67 \ (x)^2} = \frac{10,000}{1.67(5)^2} = \frac{10,000}{41.75} = 240 \text{ trees per hectare}$$

or one tree per 42 ft.^{2} The 1.67 in the denominator is a constant correction factor that has been determined by experience with this method, and x is the average distance in meters. If the measurements are made in feet, the numerator should be 43,560 instead of 10,000. This is the number of square feet in an acre, and density will be in trees per acre.

If thirty of the trees identified were sugar maple (SM), the relative density (R.D.) of this species can be determined by the following formula.

$$\text{R.D.} = \frac{\text{number of SM trees}}{\text{total number of trees}} \times \text{density of all trees} = \frac{30}{100} \times 240$$

= 72 SM trees per hectare or one sugar maple tree per 139 square meters.

After basal area for each tree has been tallied, the relative coverage for each species can be calculated. For example, if the average basal area for all sugar maple trees is 702 square centimeters and the basal area for all trees is

170,000 square centimeters, the relative coverage (R.Cv.) of sugar maple can be estimated with the following formula.

$$R.Cv. = \frac{\text{average density of SM} \times \text{average basal area of SM}}{\text{basal area of all trees}} \times 100$$

$$= \frac{72 \times 702}{170,000} \times 100 = 29.7\%$$

Sugar maple covers 29.7 percent of this forest.

If one does not wish to calculate basal areas, a very rough estimation of relative composition (R.Cm.)of this forest can be determined for each species as a percent by the following formula.

$$R.Cm. = \frac{\text{number of SM trees}}{\text{Total number of trees}} \times 100 = \frac{30}{100} \times 100 = 30\%$$

C. Why Bother?

A naturalist who wishes only to enjoy the peace and beauty of a stroll through a forest may ask, "Why bother?" This is a valid question, and the answer is that, enjoyable as a stroll through a forest may be, the sampling techniques are for those who wish to go a step further. Information gained by sampling vegetation can be useful in several ways.

1. It can be used to compare two woodlands. There will be differences in the vegetation of two areas in the same general vicinity as well as in two areas that are far removed from one another.

2. It can be used to identify changes in forest vegetation over a period of time. Change may result from the natural process of ecological succession or from the activities of humans such as logging. Fires started by human activities or by natural processes can cause great changes in vegetation. Sampling the forest vegetation before and after these events can yield insights into the way ecological succession takes place.

3. Sampling data can be used to compare forest vegetation growing under different environmental conditions. These may include such factors

as soils, precipitation, topography (north-, south-, east-, and west-facing slopes), geography (different longitudes or latitudes), and so on.

4. Professional foresters use sampling techniques to determine the number of board feet and thus the commercial value of a woodlot.

5. Information from sampling can be useful simply for the satisfaction it gives from knowing the characteristics of the vegetation. It can add depth of understanding to the peace and beauty of a stroll through a woodland.

6. Making an Ecological Calendar

The ecological calendar is based on phenological observations. Phenology is the study of the relationship of life-cycle events to weather and climate. The ecological calendar shown here is a general one with wide ranges of dates for the beginnings of seasons. It is not designed as a guide for a specific area but rather shows the range of dates for most of North America. The calendar will be different in different places because phenological events are generally later when one travels from south to north, east to west, and to higher elevations. You can construct an ecological calendar specific for your area by observing the woods around you (see chapter 6). The chief seasonal markers are events in the life cycles of deciduous trees, but some of these events usually occur naturally even in evergreen forests.

Traditional and Ecological Calendars

Traditional Calendar

Season	Date	Length
Winter	December 21-March 21	90 days
Spring	March 21-June 22	93 days
Summer	June 22-September 21	93 days
Autumn	September 21-December 21	90 days

Ecological Calendar

Marker	Starting Date	Average Length
Fall of leaves	October 31-November 30	110–166 days
Buds swell	March 20-April 15	25–42 days
Closure of canopy	May 10-May 31	86–118 days
Leaf coloration begins	August 25-September 20	72–97 days

A. The Beginning of Autumn: Autumnal Coloration

You must first decide upon the criteria you will use to identify the onset of autumnal coloration. If one leaf turns yellow, this is probably not enough evidence because it may be the result of insect damage. If fifty to one hundred leaves turn yellow, you may be justified in calling this the beginning of autumn. Leaf coloration usually appears on many trees, but sometimes a single tree may be the harbinger of things to come: even though no color is apparent in other trees, the chemical processes of autumn are already underway in them.

Another method of identifying the beginning of autumn is to observe a woodland at regular intervals and record the date when the first hint of color appears in the foliage. Whatever criteria are used, it is important to be consistent in order to make valid comparisons with succeeding years.

B. The Beginning of Winter: Leaf Fall

In a few species some scattered leaves may hang on throughout winter, but it is fairly easy to judge when most of the leaves are down. Very often leaf fall is complete after a cool night of wind and rain.

C. The Beginning of Spring: Bursting Buds

This date can be determined by observing the buds on an individual tree. When the buds begin to swell for one tree, it is an indication that the requirements for breaking dormancy have probably been met for all members of that species. Most of the deciduous species in a given area of observation break dormancy at about the same time. Although they may have different cooling and warming requirements, evergreen trees also break dormancy and produce a new crop of leaves each year.

Another way to determine the date for the beginning of spring is to observe the woodland from a distance. When a very faint tinge of green can be seen on the trees, new growth and ecological spring has begun.

D. The Beginning of Summer: Closure of the Canopy

The best way to identify this event is to go into a forest or stand under a tree and look up. As spring advances, patches of sky will be replaced with green. When this process has finished, leaves have achieved their maximum size.

E. Other Phenological Observations to Add to Your Calendar

Time of flowering. Even though they may break dormancy at about the same time, tree species may have different blooming dates. For most wind-pollinated trees, flowering time is soon after growth begins in early spring before the leaves have developed enough to interfere with pollen dispersal. This is also "hay fever" time for those who are allergic to tree pollen. In looking for flowers, it is well to remember that wind-pollinated trees do not have large showy flowers.

When twigs begin to elongate. This may coincide with or occur shortly after the buds burst. The embryonic leaves can be seen emerging from the bud. Some of the bud scales have enlarged but are still present.

When twigs stop elongating. If no elongation takes place on two consecutive observations, it can be assumed that the twig has reached maximum growth. At this time, the buds for next summer's growth are forming or are already formed. Since bud formation is genetically controlled, it will be interesting to see if each bud has the same message with regard to leaf production. This can be determined by counting the leaves on several twigs. By counting leaf scars, it can be determined if the terminal bud produced the same number of leaves last year.

When maximum leaf coloration is achieved. This is a judgment call but usually an easy one to make. The brightness of autumnal coloration is a function of the weather. If there is an abundance of precipitation with warm sunny days and cool nights, the coloration will be brilliant. If the autumn is dry, the colors will be dull, dominated by shades of the brown, dead leaves.

F. Photographing the Ecological Seasons

A visual record can be made by taking a photograph of the same scene for each of the seasons. The photograph for spring should show the green tinge of spring on the woodland followed by the full green of summer. The autumn photograph should be taken at maximum coloration, then the stark barren trees of winter. With each photograph taken at the time of day for optimal light, these four scenes can make an excellent study of contrast and color suitable for decorating any wall.

G. Comparing the Dates of Your Seasons
   with Those Farther North or South

Geographers, map makers, and navigators measure north-south distances on the earth in degrees of latitude. You can think of latitude as imaginary lines that encircle the globe parallel to the equator. They are usually marked on the margins of maps in road atlases. The lines that mark each degree of latitude are about seventy miles apart. Phenological events are about four days earlier for each degree of latitude south and four days later for each degree north. If you know the latitude of or the north-south distance between two locations, you can determine the approximate difference in time of occurrence of their ecological seasons.

   For example, Syracuse, New York, is at about 43° north latitude and Richmond, Virginia is at about 37.5° north latitude, a difference of 5.5°. Thus, ecological spring in Richmond occurs about three weeks (4 X 5.5 = 22 days) earlier than it does in Syracuse. The same conclusion can be drawn by using the north-south distance. Richmond is about 385 air miles south of Syracuse. This yields 5.5° when divided by seventy.

7. Making Leaf Collections

Many plants, especially woody plants can be identified by leaf characteristics alone. Leaf collections can thus serve as aids for identification. Waxed leaf collections can also be used for decorations. Alternate methods of making leaf collections are leaf prints and leaf skeletons. Leaf prints give an outline of the main veins of the leaf and leaf skeletons show the entire lacy network of the internal vascular system. See chapter 8 for details on preparing leaf, leaf print, and leaf skeleton collections.

8. Food for Thought

Let us suppose an imaginary nail is driven into a tree (driving real nails into trees is not a good practice) four feet above the ground with four inches protruding. If the tree grows in diameter at the rate of one-half inch per year and in height at the rate of two feet per year, how far above the ground will the nail be, and how much will be protruding in five years? Which of the following statements would you choose as correct?

1. Ten feet above the ground with one and one-half inches protruding.

2. Ten feet above the ground with two and three-fourths inches protruding.

3. Four feet above the ground with two and three-fourths inches protruding.

4. Four feet above the ground with one and one-half inches protruding.

If you chose 3, that is the correct choice. Remember a tree grows in height from the apical meristem at the tips of branches only. The tree grows in diameter by the cellular division of the vascular cambium. If the total increase in diameter in one year is one-half an inch, the increase at any one spot is one-fourth of an inch. The nail will remain at four feet and will eventually be covered completely.

Glossary

Bibliography and Further Reading

Index

Glossary

adaptation: A characteristic of an organism that contributes to its survival under the conditions of the environment.

aeration: The process of adding air.

alternate leaf: A leaf arrangement in which there is one leaf at each node.

annual plant: A plant that completes its life cycle in one year and then dies.

anther: The part of the stamen that produces pollen.

axil: The angle between the leaf and the stem.

biennial plant: A plant that lives for two years and produces its flowers and seeds in the second year.

biomass: The total amount of organic matter produced by a plant or in a given area.

blade: The flat expanded portion of a leaf.

calyx: The sepals collectively.

canopy: The continuous cover over the forest floor formed by the crowns of the tallest trees.

climax vegetation: The final stages in ecological succession composed of species that can reproduce themselves rather than being replaced by other species.

clone: A plant that is genetically identical to its parent plant.

compound leaf: A leaf in which the blade is divided into leaflets.

corolla: The petals of a flower.

cuticle: A waxy covering on all the aboveground parts of a plant.

deciduous plants: Plants that lose their leaves at the end of the growing season, as opposed to evergreen plants.

diploid: A condition in which cells contain two full sets of chromosomes, one set from the egg and one from the sperm. Zygotes and sporophytes normally are diploid.

disk flower: A tiny flower on the central disk in the flower head of the aster family, as distinct from ray flowers.

dissected leaf: A condition in which the leaf is divided into many narrow segments, as in some ferns.

ecological succession: The natural replacement of one plant community by another culminating in climax vegetation.

ecosystem: A community of living things and all the physical factors that make up the environment.

fertile: Capable of sexual reproduction.

fertilization: The union of two haploid gametes, resulting in a diploid zygote.

frond: The leaf of a fern.

gamete: A haploid sex cell such as an egg or a sperm.

gametophyte: A haploid gamete-producing structure or plant.

genus (pl. **genera**): A group of closely related plants with a common ancestor. The first word of the two-word scientific name.

germinate: To resume growth, as a seed or a dormant spore or zygote.

girdle: To remove a ring of bark around the trunk of a tree.

groundwater: The water in the ground in the saturated zone or below the water table.

habitat: The environment of an organism or a community.

haploid: Having only one set of chromosomes, as in gametes, spores, and gametophytes.

herb: A nonwoody plant that dies back to the ground at the end of the growing season; plants used in medicine or for seasoning.

herbaceous: Having the characteristics of a herb; green; having the texture of leaves, with nonwoody tissue.

herbalist: One who collects, sells, or prescribes medicinal herbs.

hydrophyte: A plant that grows in a wet environment where it is partially or completely submerged.

internode: The portion of the stem where no leaves are attached; the space between nodes.

intertidal zone: That part of the coast between low tide and high tide.

leaflet: One of the divisions that make up a compound leaf.

mesophyte: A plant that grows in environmental conditions that are intermediate with regard to moisture; between hydrophytic and xerophytic.

micelle: A very tiny soil particle.

morbid: Unnatural; not sound or healthy; diseased.

nectar: A sweet fluid produced by flowers to attract pollinators.

node: The location on a stem where one or more leaves are attached.

opposite leaves: A leaf arrangement with two leaves per node; leaves attached in pairs.

organic matter: Living or once living tissue; carbon compounds formed by living things.

ovary (plant): The enlarged basal portion of the pistil that contains the ovules and develops into the fruit.

ovule: An embryonic structure inside the ovary that will become a seed.

palmate: In compound leaves, an arrangement in which leaflets are attached at one point and radiate outward as the fingers from the palm of the hand.

perennial plant: A plant that lives for more than two years; not annual or biennial.

petals: The colorful segments of flowers that attract pollinators.

petiole: The stalk of a leaf.

phyte: A suffix that means plant, usually preceded by a descriptive prefix such as hydrophyte, xerophyte, gametophyte.

pinnate: A leaf form in compound leaves in which the leaflets are attached to each side of a central midrib.

pioneer species: The first plants to colonize bare soil or rock.

pistil: The female reproductive part of a flower; the seed-bearing part, consisting of a style, stigma, and ovary.

plant community: All the plant species growing in an area.

pollination: The transfer of pollen from the anther to the stigma.

potherb: A herbaceous plant that is edible when cooked, including the leaves and sometimes the stem.

radial: Spreading outward from a central point.

ray flower: A marginal strap-shaped flower of the aster family.

rhizome: A creeping, horizontal, underground stem.

salinity: The degree of saltiness.

sepals: The outermost parts of the flower, usually green and leaflike, which cover the outer parts of the bud.

shrub: A woody perennial not as large as a tree, usually with more than one stem.

simple leaf: A leaf that has a blade not divided into leaflets.

sp.: An abbreviation that follows the name of a genus and indicates a single unnamed or unknown species; *Acer sp.*

species: A group of organisms that can interbreed with one another but not with members of other species.

sporophyte: A diploid plant that produces haploid spores in plants that have alternation of generations.

spp.: An abbreviation that follows the name of a genus and indicates more than one unnamed or unknown species.

stamen: The male or pollen-producing structure of a flower consisting of an anther and a filament.

stigma: The part of the pistil that receives pollen and where the pollen germinates.

style: Usually a slender stalk with the stigma at one end and attached to the ovary at the other.

subspecies: A geographical race of a species.

substrate: Foundation material that makes up a given area of the earth. For example, a bog has an organic substrate.

succession: See **ecological succession.**

succulent: Thick, juicy, fleshy, as in the leaves and stems of plants adapted for dry environments.

summergreen: A term sometimes used to describe the eastern deciduous forest that is green in the summer only, as opposed to evergreen.

terrestrial: A land plant, as opposed to aquatic.

thallus: A plant body that is not modified into root, stem, and leaf, as in some of the liverworts.

transpiration: The loss of water by evaporation from the surface of plants.

understory trees: Trees that grow beneath the canopy of a forest but do not become part of the canopy.

vegetation: The sum of all the plants.

viable: Alive and capable of growth, as a seed.

water table: The top surface of the groundwater.

whorled leaves: An arrangement of leaves with three or more attached at a node.

windfall: Trees blown down by the wind.

wort: A suffix that means plant.

xerophyte: A plant adapted to live under dry conditions.

zygote: A diploid cell formed by the union of two haploid gametes.

Bibliography and Further Reading

Anderson, Frank J. 1997. *An Illustrated History of the Herbals*. New York: Columbia University Press.

Bailey, Liberty Hyde. 1933. *How Plants Get Their Names*. New York: Macmillan Co.

Barbour, M. G., J. H. Burk, and W. D. Pitts. 1980. *Terrestrial Plant Ecology*. Menlo Park, Calif.: Benjamin/Cummings Publishing Co., Inc.

Bell, C. Richie, and B. J. Taylor. 1982. *Florida Wild Flowers*. Chapel Hill, N.C.: Laurel Hill Press.

Benson, Lyman. 1979. *Plant Classification*. Lexington, Mass.: D. C. Heath and Co.

Berlin, Brent. 1973. "Folk Systematics in Relation to Biological Classification and Nomenclature." *Annual Review of Ecology and Systematics*, vol. 4. Palo Alto, Calif.: Annual Reviews, Inc.

Billings, W. D. 1964. *Plants and the Ecosystem*. Belmont, Calif.: Wadsworth Publishing Co., Inc.

Braun, E. Lucy. 1950. *Deciduous Forest of Eastern North America*. New York: Macmillan Publishing Co., Inc.

Brayshaw, T. Christopher. 1996. *Plant Collecting for the Amateur*. Victoria, British Columbia: Royal British Columbia Museum.

Brown, Lester R., N. Lenssen, and H. Kane. 1995. "Tropical Forests are Vanishing." In *Vital Signs 1995*, edited by Linda Starke, 116–17. New York: W. W. Norton and Co.

Brown, Lester, et. al. 1990. *State of the World 1990*. New York: W. W. Norton and Co.

Buchholz, Rogane A. 1998. *Principles of Environmental Management*. Upper Saddle River, N.J.: Prentice-Hall.

Campbell, F. T. 1980. "Conserving Our Wild Plant Heritage." *Environment* 22 (9): 14–20.

Carlson, Eric, D. Cusick, and C. Taylor. 1992. *The Complete Book of Nature Crafts.* Emmaus, Penn.: Rodale Press.

Charas, Daniel D. 1992. *Lessons From Nature.* Washington, D.C.: Island Press.

Coffey, Timothy. 1993. *The History and Folklore of North American Wildflowers.* New York: Facts on File, Inc.

Courtenay, B., and H. H. Burdsall Jr. 1984. *A Field Guide to Mushrooms and Their Relatives.* New York: Van Nostrand Reinhold.

Cox, Donald D. 1996. *Seaway Trail Wildguide to Natural History.* Sackets Harbor, N.Y.: Seaway Trail Foundation.

Cox, Donald D. 1985. *Common Flowering Plants of the Northeast.* Albany: State University of New York Press.

Crawley, M. J., ed. 1986. *Plant Ecology.* Boston: Blackwell Scientific Publications.

Croom, Edward M. 1983. "Documenting and Evaluating Hebal Remedies." *Economic Botany* 37 (1): 13–27.

Cutter, Susan L., H. L. Renwick, and W. H. Renwick. 1985. *Exploitation, Conservation, Preservation.* Totowa, N.J.: Rowman and Allanheld Publishers.

Daubenmire, R. F. 1959. *Plants and Environment.* 2d ed. New York: John Wiley and Sons, Inc.

Davis, Richard C., ed. 1983. *Encyclopedia of American Forest and Conservation History,* 2 vols. New York: Macmillan Publishing Co.

Eckholm, Erik P. 1976. *Losing Ground, Environmental Stress, and World Food Prospects.* New York: W. W. Norton and Co., Inc.

Ehrlich, Paul, and Anne Ehrlich. 1981. *Extinction: The Causes and Consequences of the Disappearance of Species.* New York: Random House.

Fahn, Abraham, and E. Werker. 1972. "Anatomical Mechanisms of Seed Dispersal." In *Seed Biology,* vol. 1, edited by T. T. Kozlowski, 151–221. New York: Academic Press.

Fernald, Merritt L. 1950. *Gray's Manual of Botany.* 8th ed. corrected printing, 1970. New York: D. Van Nostrand Co.

Gibbons, Euell. 1966. *Stalking the Heathful Herbs.* Brattleboro, Vt.: Alan C. Hood and Co., Inc.

Gibbons, Euell, and G. Tucker. 1979. *Euell Gibbons Handbook of Edible Wild Plants.* Virginia Beach, Va.: Unilaw Library Press.

Given, David R. 1994. *Principles and Practice of Plant Conservation.* Portland, Ore.: Timber Press.

Gleason, Henry A., and A. Cronquist. 1991. *Manual of Vascular Plants of Northeastern United States.* New York: New York Botanical Garden.

Gleason, H. A., and A. Cronquist. 1964. *The Geography of Plants.* New York: Columbia University Press.

Hale, M. E. 1979. *How to Know Lichens.* 2d ed. Dubuque, Ia.: Wm. C. Brown Co.

Hardin, James W., and J. M. Arena. 1974. *Human Poisoning from Native and Cultivated Plants*. Durham, N.C.: Duke University Press.

Harper, J. L., P. H. Lovell, and K. G. Moore. 1970. "The Shapes and Sizes of Seeds." In *Annual Review of Ecology and Systematics*, vol. 1, edited by R. F. Johnston, P. W. Frank, and C. D. Michener, 327–56. Palo Alto, Calif.: Annual Reviews, Inc.

Hitchcock, S. T. 1980. *Gather Ye Wild Things*. New York: Harper and Row.

Holbrook, Stewart T. 1944. *Burning an Empire*. New York: Macmillan Publishing Co.

Howe, Henry F., and J. Smallwood. 1982. "Ecology of Seed Dispersal." In *Annual Review of Ecology and Systematics*, vol. 13, edited by R. F. Johnston, P. W. Frank, and C. D. Michener, 201–28. Palo Alto, Calif.: Annual Reviews Inc.

Jaeger, Edmond C. 1955. *The California Deserts*. Stanford, Calif.: Stanford University Press.

Johnson, A. S., H. O. Hillestad, S. F. Shanholtzer, and G. F. Shanholtzer. 1974. *An Ecological Survey of the Coastal Region of Georgia*. Washington, D.C.: U.S. Government Printing Office.

Joosten, Titia. 1988. *Flower Drying with a Microwave: Techniques and Projects*. New York: Sterling Publishing Co., Inc.

Kaufman, Peter B., T. F. Carson, P. Dayanandan, M. L. Evans, J. B. Fisher, C. Parks, and J. R. Wells. 1991. *Plants: Their Biology and Importance*. 2d ed. Philadelphia: Harper and Row Publishers.

Keeney, Elizabeth B. 1992. *The Botanizers*. Chapel Hill: University of North Carolina Press.

Kent, Donald M. 1994. *Applied Wetlands Science and Technology*. Boca Raton, Fla.: Lewis Publishers.

Ketchledge, E. H. 1970. *Plant Collecting: A Guide to the Preparation of a Plant Collection*. Syracuse: State University of New York College of Environmental Science and Forestry.

Kinghorn, A. Douglas. 1979. *Toxic Plants*. New York: Columbia University Press.

Kingsbury, John M. 1964. *Poisonous Plants of the United States and Canada*. Englewood Cliffs, N.J.: Prentice-Hall, Inc.

Knobloch, Irving W. 1963. *Selected Botanical Papers*. Englewood Cliffs, N.J.: Prentice-Hall, Inc.

Koopowitz, Harold, and Hilary Kaye. 1983. *Plant Extinction: A Global Crises*. Washington, D.C.: Stone Wall Press, Inc.

Kowalchik, Claire, and W. H. Hylton, eds. 1987. *Rodale's Illustrated Encyclopedia of Herbs*. Emmaus, Penn.: Rodale Press.

Krochmal, Connie, and Arnold Krochmal. 1973. *A Guide to the Medicinal Plants of the United States*. New York: Quadrangle/New York Times Book Co.

Lampe, Kenneth F., and M. A. McCann. 1985. *AMA Handbook of Poisonous and Injurious Plants*. Chicago, Ill.: American Medical Association.

Leopold, Aldo. 1966. *A Sand County Almanac*. New York: Ballantine Books.

Lewis, W. H., and M. Elvin-Lewis. 1977. *Medical Botany: Plants Affecting Man's Health*. New York: John Wiley and Sons.

Lincoff, G. H. 1981. *The Audubon Society Field Guide to North America Mushrooms*. New York: Alfred A. Knopf.

Litovitz, Toby L., L. R. Clark, and R. A. Solway. 1993. *Annual Report of the American Association of Poison Control Centers*. Washington, D.C.

Lyman, Francesca, I. Mintzer, K. Courrier, and J. Mackenzie. 1990. *The Greenhouse Trap*. Boston: Beacon Press.

MacFarlane, R. B. 1985. *Collecting and Preserving Plants for Science and Pleasure*. New York: Arco Publishing Inc.

Meeuse, B. J. D. 1961. *The Story of Pollination*. New York: Ronald Press Co.

Merchand, Peter J. 1987. *Life in the Cold: An Introduction to Winter Ecology*. Hanover, Mass.: University Press of New England.

Miller, David F., and G. W. Blades. 1962. *Methods and Materials for Teaching the Biological Sciences*. 2d ed. New York: McGraw-Hill Book Co., Inc.

Miller, G. Tyler, Jr. 1992. *Living in the Environment*. Belmont, Calif.: Wadsworth Publishing Co.

Millspaugh, Charles F. 1974. *American Medical Plants*. New York: Dover Publications, Inc.

Morton, A. G. 1981. *History of Botanical Science*. New York: Academic Press.

Munzer, Martha E., and P. F. Brandwein. 1960. *Teaching Science Through Conservation*. New York: McGraw-Hill Book Co., Inc.

Myers, Norman. 1984. *The Primary Source*. New York: W. W. Norton and Co.

Niehaus, T. F., and C. L. Ripper. 1976. *Pacific States Wildflowers*. Boston: Houghton Mifflin Company.

Niering, William, and N. Olmstead. 1979. *Audubon Society Field Guide to North American Wildflowers (Eastern Region)*. New York: Alfred A. Knopf Co.

Perlin, John. 1991. *A Forest Journey*. Cambridge, Mass.: Harvard University Press.

Peterson, Lee. 1977. *A Field Guide to Edible Wild Plants*. Boston: Houghton Mifflin Co.

Richardson, W. Norman, and T. H. Stubbs. 1976. *Evolution, Human Ecology, and Society*. New York: Macmillan Publishing Co., Inc.

Saunders, C. F. 1948. *Edible and Useful Wild Plants of the United States and Canada*. New York: Dover Publications, Inc.

Schery, Robert W. 1972. *Plants for Man*. 2d ed. Englewood Cliffs, N.J.: Prentice-Hall, Inc.

Sears, Paul B. 1966. *The Living Landscape*. New York: Basic Books, Inc.

Simpson, Beryl Brintnall, and M. Conner-Ogorzaly. 1986. *Economic Botany: Plants in Our World*. New York: McGraw-Hill Co.

Small, John Kunkel. 1972. *Manual of Southeastern Flora*. New York: Hefner Publishing.

Stebbins, G. Ledyard. 1971. "Adaptive Radiation of Reproductive Characteristics in Angiosperms II: Seeds and Seedlings." In *Annual Review of Ecology and Systematics*, vol. 2, edited by R. F. Johnston, P. W. Frank, and C. D. Michener, 237–60. Palo Alto, Calif.: Annual Reviews, Inc.

Tippo, Oswald, and W. L. Stern. 1977. *Humanistic Botany*. New York: W. W. Norton and Co., Inc.

Turner, Nancy J., and A. F. Szczawinski. 1991. *Common Poisonous Plants and Mushrooms of North America*. Portland, Ore.: Timber Press.

Van der Pijl, L. 1972. *Principles of Dispersal in Higher Plants*. 2d ed. New York: Springer-Verlag.

Index